Monographs in Electrical and Electronic Engineering 31

Series editors: P. Hammond, T.J.E. Miller, and S. Yamamura

Monographs in Electrical and Electronic Engineering

Switched Reluctance Motors and Their Control

TJE Miller
Lucas Professor in Power Electronics
SPEED Laboratory
University of Glasgow

MAGNA PHYSICS PUBLISHING
AND
CLARENDON PRESS · OXFORD
1993

Magna Physics Publishing, Hillsboro, OH 45133
Magna Physics Div. Tridelta Industries Inc., Mentor, OH 44060
and
Oxford University Press, Walton Street, Oxford OX2 6DP

Oxford   New York   Toronto
Delhi   Bombay   Calcutta   Madras   Karachi
Kuala Lumpur   Singapore   Hong Kong   Tokyo
Nairobi   Dar es Salaam   Cape Town
Melbourne   Auckland   Madrid

and associated companies in
Berlin   Ibadan

Oxford is a trade mark of Oxford University Press

Published in the United States
by Magna Physics Div., Tridelta Industries Inc., Hillsboro, Ohio
and Oxford University Press Inc., New York

Library of Congress Cataloging in Publication Data is Available

A catalogue record for this book is available from the British Library

Magna Physics ISBN 1-881855-02-3
OUP ISBN 0-19-859387-2

Typeset by the Author

Printed in the United States of America
by Book Crafters, Chelsea, Michigan 48118-0370

Preface

We have produced this book to help engineers understand, design and evaluate the switched reluctance motor and its control. The electromagnetic principle of the motor is quite simple, but it is nonlinear and this makes it difficult to develop accurate yet simple design formulas. From the designer's point of view, it is a classic case where "pushing the balloon in one place makes it bulge out somewhere else"[1]. In this sense it is no different from any other engineering design problem. We have tried to indicate the best places to push the balloon, and where to look out for bulging.

Evaluation of the switched reluctance motor is important because it is a relative newcomer to the stable of electric motors. We have tried to give the reader the tools and some of the information needed to make his or her own judgement, but we have included pictorial evidence of some interesting applications developed mainly by others (to whom we express our thanks and appreciation).

The switched reluctance motor is a fascinating machine with all kinds of interesting application possibilities, and we hope that the reader will enjoy this account of it.

The design procedures described here have been used successfully for several years in the design and testing of prototype switched reluctance motors for industrial companies by the Glasgow University SPEED Consortium and by Magna Physics Corporation, Hillsboro, Ohio. The book embraces much of the thinking behind the computer program *PC-SRD*, which permits rapid preliminary design of both motor and controller. It includes know-how accumulated over many years from the continual interaction between the software-development and hardware test laboratories on the one hand, and industrial designers and users on the other. It is also a product of

---

[1]This expression is attributed to Dr. Eike Richter.

collaboration over many years between the author and publisher.

When the switched reluctance motor first became widely known in the early 1980's the technology was shrouded in mystery. Claims were made for its performance and operating characteristics that seemed hardly believable at the time. Design procedures were not publicly available, and companies had no means of verifying claims or of developing opportunities themselves. So there began the slow process of testing whatever prototypes could be purchased or quickly fabricated, and the even slower process of developing validated design techniques and software.

Part of the design problem is that the switched reluctance motor does not conform to the classical design techniques used for classical DC and AC electric motors. The practical design engineer is confronted with a machine that has no steady state, has extreme localized saturation, and requires an unfamiliar power-electronic converter to make it work at all. The geometry is beguilingly simple, and everything about the motor and its control seems *at first sight* to be a gift to the production engineer. Yet the attainment of good designs and satisfactory performance is practically impossible by traditional design methods.

The switched reluctance motor *does* obey the laws of physics and *can* be designed by any competent team with the necessary electromagnetic, mechanical, and electronic engineering skills. The biggest hurdle is the initial one of knowing how to start. This book lays out an organized design approach to deal with the basic design calculations, together with a collection of relevant data and examples. From this structure, engineers can develop particular designs to meet their needs.

Most of the difficulty in understanding the operation and design of the switched reluctance motor stems from the *double saliency*: neither the rotor nor the stator has a smooth cylindrical surface at the airgap, but both members have salient poles or teeth. Because of this feature, there is no pure steady-state. "Steady-state" operation is a series of transients in which the inductive circuit parameters of the phase windings are variable functions of both position and current. The variation with current is associated with intense saturation of partially-overlapping poles, which is difficult to calculate in detail without finite-element software.

This means that the computer is essential for design. However, the fundamental physics remains relatively simple, so that while the numbers must be worked out by the computer, their meaning can be readily interpreted physically. This book deals with both aspects: the design algorithms *and* their physical interpretation. The design principles are close to those used in *PC-SRD*, but readers do not need to own *PC-SRD* to understand them. Throughout, design decisions are reinforced by straightforward engineering principles that may appear surprisingly simple to anyone expecting profound and obscure theories.

Many years ago the same process was followed in the development of design procedures for the induction motor and the DC commutator motor. We believe that this book is the most complete account of switched reluctance motor design yet published. It includes practical formulas never published before, and assembles them for the first time with many others into a coherent design procedure.

Inevitably, with the pace of development, there are many recent ideas and improvements taking place that have not found their way into this book, and for these we can do no more than point to the *IEE* and *IEEE* journals and relevant conference publications.

TJE Miller   JR Hendershot Jr
Glasgow, 1992   Hillsboro, 1992

Acknowledgements

Thanks are due to the many engineers who have participated in the development of prototype switched reluctance motors using *PC-SRD* and to those who have worked with and supported the SPEED laboratory at Glasgow University. Special acknowledgement is due to Jeff Coles and Mel Ward (Lucas Automotive), Gary Horst (Emerson), J.H. Johnson and Dave Thimmesch (A.O. Smith), Dave Anderson (National Semiconductor UK Ltd.), and Alex Krinickas (Sundstrand). Particular acknowledgement is also due to Professor Martyn Harris of the University of Southampton for innumerable discussions over many years about the nature and properties of switched reluctance motors and the means for analysing, designing, and testing them. Also to the research assistants in the SPEED laboratory who have contributed in the software and testing areas, particularly Malcolm McGilp, Calum Cossar, Dave Staton, Gary Gray and Alan Hutton (now with Motorola UK); technicians Ian Young and Peter Miller; and research students W.L. Soong and Rolf Lagerquist. TJEM would also thank his ex-colleagues at General Electric's Research and Development and at GE Motors, where much of the leading development work has taken place in the United States. Special acknowledgment is also due to Professor P.J. Lawrenson, originally of the University of Leeds and now of SR Drives Ltd, who initiated the "crusade" to develop the SR motor and who has been supportive in a number of ways; and to several of his colleagues, together with companies who have provided examples of SR products developed under licence to SRDL over a wide range of applications. Individual acknowledgments occur in the text but we would specially like to thank Peter McCreath of Allenwest, Paul Greenough of British Jeffrey Diamond (a Division of the Dresser company), John Holden of HIL Electric Vehicles, and Colin Shiers of Graseby Controls.

Contents

Switched Reluctance Motors and their Control

1. INTRODUCTION

1.1 Definition. A *reluctance motor* is an electric motor in which torque is produced by the tendency of its moveable part to move to a position where the inductance of the excited winding is maximized.

The motion may be rotary or linear, and the rotor may be interior (as in Fig. 1.1) or exterior. The "winding" usually consists of a number of electrically separate circuits or *phases*. These may be excited separately or together. In motoring operation, each phase is usually excited when its inductance is increasing, and is unexcited when its inductance is decreasing. In generating, the opposite is true.

This definition is broad enough to include both the *switched* reluctance motor and the *synchronous* reluctance motor, Fig. 1.2. The *idealised forms* of these machines are defined as follows.

1.2 Idealised switched and synchronous reluctance motors

Switched Reluctance

1. Both stator and rotor have salient poles.

2. The stator winding comprises a set of coils, each of which is wound on one pole.

3. Excitation is a sequence of current pulses applied to each phase in turn.

4. As the rotor rotates, the phase *flux-linkage* should have a triangular or sawtooth waveform but not vary with current.

Synchronous reluctance

1. The stator has a smooth bore except for slotting.

2. The stator has a polyphase winding with approximately sine-distributed coils.

3. Excitation is a set of polyphase balanced sinewave currents.

4. The phase self-inductance should vary sinusoidally with rotor position but not vary with current.

The significance of items 4 will not be apparent until the torque-production process is discussed in later chapters.

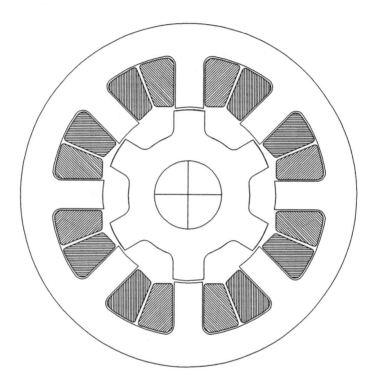

Fig. 1.1   Switched reluctance motor with 8 stator poles and 6 rotor poles. Each phase winding comprises two coils, wound on opposite poles.

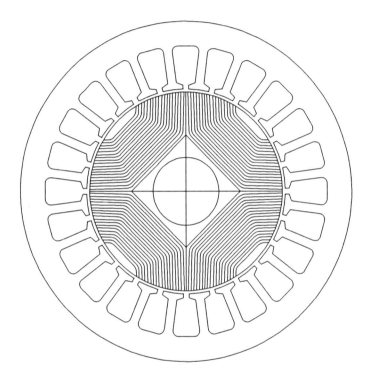

Fig. 1.2   Axially-laminated synchronous reluctance motor. This
              is a true AC motor. The stator is essentially the same as
              that of the induction motor, and the supply is sinusoidal.

1.3 Other terms used for switched reluctance motors. In the United States the term *variable reluctance motor* (VR motor) is often preferred, even though the term *switched reluctance* originated there: it was used by S.A. Nasar to describe a rudimentary switched reluctance motor in 1969 [1]. However, the VR motor is also a form of stepper motor, so this term is liable to cause confusion (see §1.4). Use of the term *switched reluctance* in connection with the modern form of SR motor is undoubtedly due to Professor Lawrenson and his colleagues at Switched Reluctance Drives Ltd. [2]. The term *brushless reluctance motor* has also been used occasionally to underline the fact that the motor is brushless [3]; and *electronically commutated reluctance* (ECR) motor has also been used [4]. The term *switched reluctance* does not mean that the reluctance itself is switched, but it clearly refers to the switching of phase currents, essential to operation. This switching is more precisely called *commutation*, so *ECR* is an even more precise term than *switched reluctance*. It also draws a parallel with the electronically commutated permanent-magnet motor (i.e. the squarewave or "trapezoidal" brushless DC motor). In both cases the main function of the switching is the same as that of the commutator in a DC motor.

1.4 Relationship with VR stepper motors. The switched reluctance motor is topologically and electromagnetically identical to the VR stepper, Fig. 1.1. The differences are in engineering design, in the control method, and in performance and application characteristics. The main differences are as follows:

Switched reluctance motor

VR stepper motor

1. Normally operated with shaft-position feedback to synchronize the commutation of the phase currents with precise rotor positions.

1. Normally run *open-loop*, i.e. without shaft-position feedback.

2. Normally designed for efficient conversion of power, up to at least 300kW.

2. Normally designed to maintain *step-integrity* rather than to achieve efficient power conversion.

1.5 History of the switched reluctance motor. The earliest recorded switched reluctance motor was the one built by Davidson in Scotland in 1838 and used to propel a locomotive on the Glasgow-Edinburgh railway near Falkirk. Dr. Fulton of SRDL recently described Davidson's locomotive in a seminar of the UK Magnetics Club, pointing out that the locomotive weighed several tons yet the top speed was less than could be achieved with one man pushing. (See [5].)

The Lucas Ledex rotary actuator dates back to World War II, and continues in production today, but this is a limited-rotation actuator rather than a motor.

The stepper motor, invented and patented in the 20's by C.L. Walker in Aberdeen, included many of the features of modern VR stepper motors and therefore of the switched reluctance motor.

Two US patents filed by Bedford and Hoft in 1971 and 1972 describe many of the essential features of the modern switched reluctance motor, with true electronic commutation positively synchronized with rotor position [6]. Bedford and Hoft discussed rotor geometry as well as the circuit topology of the power-electronic controller. However, this cannot be regarded as a master patent because the same electromagnetic and control principles were used on Davidson's machine.

Important milestones in the recent history include the axial-gap, thyristor-controlled motor built by Unnewehr and Koch of Ford Motor Company [7], and other works by Bausch [8].

In Europe the commercial potential of the switched reluctance motor was realised by Byrne [9] and Lawrenson [2], but the rapid exploitation and technical development by SRDL probably played the leading role in exciting interest in the technology especially in the early 80's. Among SRDL's early licensees, the best known is *Tasc Drives* (now *Graseby Controls*) who manufacture a range of general-purpose variable-speed switched-reluctance drives for industrial applications. These cast-iron framed motors have explosion-proof certification in the UK, and cover the range from 4kW to 80kW.

More recently products have been announced by *Allenwest Electrical*,

Prestwick, Scotland (the *Motionmaster* range of industrial cast-iron-framed motors from 4kW to 75kW, with Toshiba IGBT phaseleg modules developed specially for switched reluctance motors); and by *British Jeffrey Diamond*, a subsidiary of Dresser (flameproof mining motors and controllers at 35kW and 150kW, 1100V). These products were displayed at the 1991 Total Solutions Exhibition at the UK National Exhibition Centre in Birmingham. *Radioenergie* in France has also announced a switched reluctance drive for low-voltage DC applications [10,11].

In the United States the first commercial application was the *Hewlett-Packard* servo drive used in the *Draftmaster* computer-plotter [13]. Konecny's description of this motor includes an account of the use of controlled saturation at the airgap to decrease the torque ripple [14]. Later work by Stephenson confirmed the theoretical validity of this principle. The Hewlett-Packard motors are controlled by a special integrated circuit (the HCTL1000, subsequently replaced by the HCTL1100) which incorporates many advanced features and can also control brushless DC motors. *Semifusion* in Santa Clara, CA, subsequently developed several switched-reluctance servo-drives using this IC.

At the time of writing, the list of established commercial applications is short. In spite of widespread development effort, and apart from a few notable exceptions, the *time to market* has so far proven to be very long. What this indicates for the future of switched reluctance motors, only time will tell.

2 Energy Conversion Principles of the Switched Reluctance Motor

2.1 Magnetization curves. Fig. 2.1 shows a 6/4 motor, i.e., one having 6 stator and 4 rotor poles. The 6/4 motor has three phases. Each phase comprises two coils wound on opposite poles and connected so that their fluxes are additive. They may be in series or in parallel, but we will assume for now that they are in series.

The aligned position

When any pair of rotor poles is exactly aligned with the stator poles of phase 1, that phase is said to be in the *aligned position*, as shown in Fig. 2.1(a). [The phase 1 poles are on the horizontal axis.] When current is flowing in phase 1, there is no torque at this position because the rotor is in a position of maximum inductance. If the rotor is displaced to either side of the aligned position, as in Figs. 2.1(c) and (d), there is a restoring torque that tends to return the rotor towards the aligned position. In stepper-motor terms, the aligned position is the *detent* position.

In the aligned position the phase inductance is at its maximum because the magnetic reluctance of the flux path is at its lowest. At low current levels most of the reluctance is in the airgap, but the long path through the stator yoke can also absorb a significant MMF and reduce the aligned inductance appreciably, even at low currents.

In the aligned position the flux-path is susceptible to saturation, especially in the stator and rotor yokes. The current at which saturation begins can be estimated by assuming that the cross-section of the flux-path is uniform. This implies that the radial depth y_s of the stator yoke is equal to one-half the stator tooth-width t_s, and likewise for the rotor. (This is what is meant by saying that the magnetic circuit has a "uniform" cross-section). If steel is assumed to saturate at a flux-density of 1.7T, the ampere-turns *per pole* required to produce this flux-density in the airgap (and throughout the uniform magnetic circuit) is given by

$$N_p i_s = \frac{B_s}{\mu_0} \cdot g \qquad (1)$$

where N_p is the turns/pole.

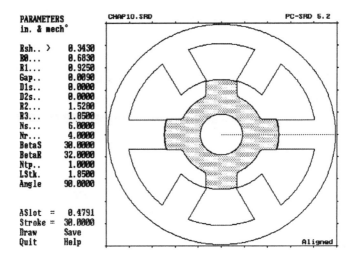

Fig. 2.1 (a)   6/4 s.r. motor - *aligned position* on phase 1. The poles of phase 1 are on the horizontal axis

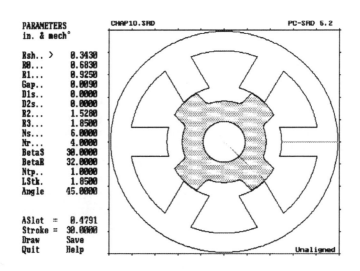

Fig. 2.1 (b)   6/4 s.r. motor - *unaligned position* on phase 1

Page 8

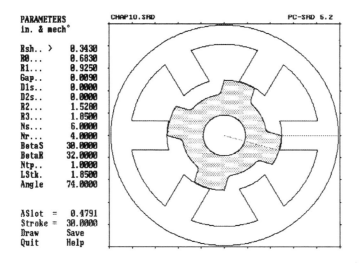

Fig. 2.1 (c) 6/4 s.r. motor - *partial overlap* position on phase 1 while motoring in the counterclockwise direction

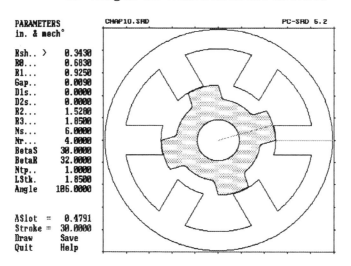

Fig. 2.1 (d) 6/4 s.r. motor - *partial overlap* position on phase 1 while *generating* in the counterclockwise direction

If g = 0.2mm (0.008"), the required m.m.f. is only 270 ampere-turns/pole. The aligned *magnetization curve* is therefore of the form shown in Fig. 2.2.

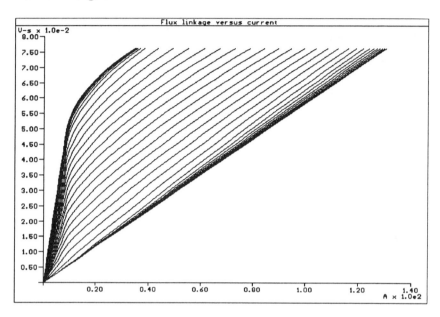

Fig. 2.2   Complete set of magnetization curves, showing flux-linkage vs. current for one phase, with the rotor at several positions between the unaligned and aligned positions. The aligned curve is the highest, and the unaligned curve is the lowest.

The unaligned position

When the *inter*polar axis of the rotor is aligned with the poles of phase 1, phase 1 is in the *unaligned position*, as shown in Fig. 2.1(b). When current is flowing in phase 1, there is no torque at this position. If the rotor is displaced to either side of the unaligned position, there appears a torque that tends to displace it still further and attract it towards the next aligned position. The unaligned position is one of unstable equilibrium.

In the unaligned position the phase inductance (the slope of the unaligned curve in Fig. 2.2) is at its minimum, because the magnetic reluctance of the flux path is at its highest as a result of the large airgap between the stator and the rotor. The airgap reluctance is much greater than that of the steel sections, in spite of the long magnetic path length through the stator yoke.

The unaligned magnetization curve is not as susceptible to saturation as is the aligned curve. The current at which saturation begins can be estimated as follows. Let L_{a0} be the unsaturated inductance in the aligned position (this is the slope of the aligned magnetization curve in the linear region). Let L_{u0} be the unsaturated inductance in the unaligned position, and let λ be the *inductance ratio* L_{a0}/L_{u0}. Then, assuming the magnetic circuit to have a uniform cross-section as before, the ampere-turns/pole needed to produce a maximum flux-density of B_s in the stator yoke is $\lambda N_p i_s$, and this will be of the order of 2700At for the earlier example. The saturation of the unaligned curve will be less sharp than that of the aligned curve, because the leakage flux is relatively much greater in the unaligned case. The two curves converge at high flux levels, but they can never intersect.

Intermediate rotor positions

At intermediate rotor positions such as the ones shown in Fig. 2.1(c) and (d), the magnetization curve is intermediate between the aligned and unaligned curves. If there is any overlap at all, the possibility exists for local saturation of the pole-corners. The *local* flux-density at which this saturation begins is B_s as before. Ampère's circuital law can be applied to a contour right through the saturated pole-corners and across the airgap, and it is clear that the ampere-turns required at the onset of local saturation is just $N_p i_s$, the same value as that required in the aligned position. Increasing the current beyond this value causes the saturated region to grow and spread, initially in the region of the overlapping pole-corners but ultimately through the entire magnetic circuit. The magnetization curves at intermediate rotor positions therefore have the form shown in Fig. 2.2.

Because of the sharpness of the pole-corners it would not be surprising if there were a sudden change in magnetization characteristics at the start of overlap, i.e. near the position shown in Fig. 2.1(c). This is indeed the case. Between the unaligned position and the start of overlap, the magnetization curves do not change very rapidly. As the start-of-overlap is approached, the curves begin to sweep upwards and rapidly assume a shape closer to that of the aligned curve. In the last few degrees before alignment, again there is little change in the curves.

The importance of the magnetization curves for the calculation of torque, and therefore for the design of the laminations and windings, cannot be overstated.

It is important to note that the *inductance* of a phase varies widely as a function of both rotor position and phase current. The most meaningful inductances in the theory of the switched reluctance motor are the unsaturated aligned inductance L_{a0} and the unsaturated unaligned inductance L_{u0}. When considering the chopping of phase current it is sometimes helpful to consider also the incremental inductance l especially in the saturated region near the aligned position. This incremental inductance can have a value significantly lower than L_{u0}, with important consequences for the design of current-regulating and/or current-limiting circuitry.

The phase inductance is also important as a circuit concept that makes the control of the switched reluctance motor more intelligible to the electronics engineer. One of the classic early control papers was written entirely in terms of phase inductance [15], negecting the effects of saturation altogether. If we define inductance as the ratio ψ/i, where ψ is the actual flux-linkage, then the form of the inductance curves as a function of rotor position, with current as a parameter, is shown in Fig. 2.3. These curves are periodic in τ, the rotor pole-pitch. If N_r is the number of rotor poles,

$$\tau = \frac{2\pi}{N_r}.$$ [2]

2.2 Instantaneous Torque. When current flows in a phase, it is axiomatic that the torque tends to move the rotor in such a direction as to increase the inductance, until it reaches the position where the inductance has a maximum value. Provided that there is no residual magnetization in the steel, the direction of current is immaterial.

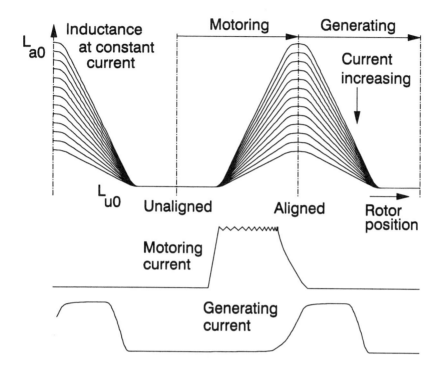

Fig. 2.3   Inductance vs. rotor position, showing the effect of saturation as the current increases. The phasing of motoring and generating current pulses is also shown.

The direction of the torque is always towards the nearest aligned position. Therefore positive torque (i.e. motoring torque) can be produced only if the rotor is between the unaligned position and the next aligned position in the forward direction. In other words, motoring torque can be produced only in the direction of rising inductance. This is illustrated in Figs. 2.3 and 2.4. If the rotor and stator poles are symmetric, each phase can produce unidirectional torque over only *half* a rotor pole-pitch. Consequently at least two phases are needed to produce unidirectional torque at all rotor positions. Fig. 2.4 shows the production of unidirectional torque in the 6/4 motor by means of overlap between the current waveforms of three phases.

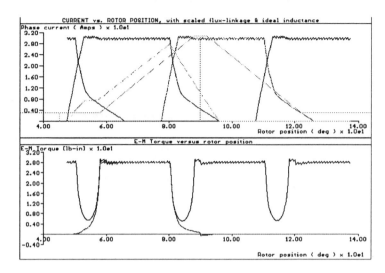

Fig. 2.4   (Upper) Three phase currents, idealised inductance, and flux-linkage of phase 1 in 6/4 motor.

(Lower) Instantaneous torque. The total torque is shown, together with the torque of phase 1.

If current is flowing when the rotor is in a position where the inductance is decreasing in the direction of rotation, the torque is negative (i.e. braking or generating).

Fig. 2.4 illustrates a difficulty with the 6/4 motor: the torque-dip that occurs near the commutation from one phase to the next. In spite of the fact that full phase current is flowing for virtually the whole time in at least one phase, there is still a dip in the torque waveform. The torque deficiency in this example is associated with the incoming phase. For high values of running torque, it can be reduced only by widening the stator and rotor poles, but this decreases the slot area and increases the copper losses. The conflict is difficult to resolve in three-phase switched reluctance motors. For smooth torque, four-phase motors are preferable. For low values of running torque, the torque dip in the three-phase motor can be reduced by *current profiling*. In effect, the phase current is boosted in regions where there is a torque dip.

The most general expression for the torque produced by one phase at any rotor position is

$$T = \left[\frac{\partial W'}{\partial \theta}\right]_{i=const.} \quad [3]$$

where W' is the *coenergy*. At any position the coenergy is the area below the magnetization curve as shown in Fig. 2.5, in other words, the definite integral

$$W' = \int_0^{i_1} \psi \, di \quad [4]$$

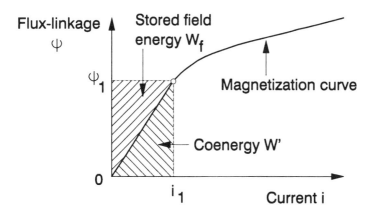

Fig. 2.5   Definition of *coenergy W'* and *stored field energy* W_f

From these equations the instantaneous torque can be visualized graphically: it is the work ΔW_m divided by $\Delta\theta$, where ΔW_m is evolved *at constant current* as the rotor moves through an infinitesimal displacement $\Delta\theta$. This is illustrated in Fig. 2.6. During such a displacement there is an exchange of energy with the supply, and there is also a change in the stored field energy. The constant-current constraint ensures that during such a displacement, the mechanical work done is exactly equal to the change in coenergy. This can be proved as follows. In a displacement $\Delta\theta$ from A to B in Fig. 2.6 at constant current, the energy exchanged with the supply is

Page 15

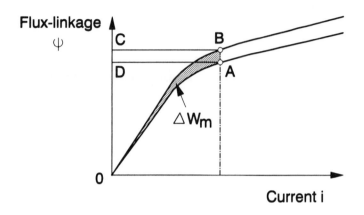

Fig. 2.6   Calculation of instantaneous torque from the rate of change of coenergy at constant current

$$\Delta W_e = ABCD. \qquad [5]$$

The change in stored field energy is

$$\Delta W_f = OBC - OAD \qquad [6]$$

and the mechanical work done must be

$$
\begin{aligned}
\Delta W_m &= T\Delta\theta \\
&= \Delta W_e - \Delta W_f \\
&= ABCD - (OBC - OAD) \\
&= (ABCD + OAD) - OBC \\
&= OAB.
\end{aligned} \qquad [7]
$$

Not all the energy obtained from the supply is converted to mechanical work. Some of it is stored in the magnetic field. The energy stored in the magnetic field is not wasted, but it is not *available* for energy conversion during the motion from A to B. This has an important effect on the rating of the controller and the need for filter capacitors, as we shall see. The effect is similar to the operation of AC motors with a lagging power factor.

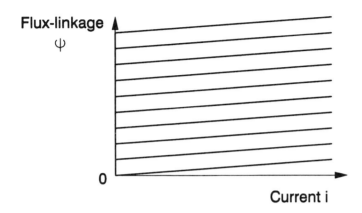

Fig. 2.7   Ideally-saturable magnetization curves, obtainable only with zero airgap and a hypothetical core material that has a perfectly bilinear B/H characteristic.

It can be seen from Fig. 2.7 that if the magnetization curves were "ideally saturable", with infinite unsaturated aligned inductance, the change in stored field energy as the rotor moves would be negligible. In this case all of the energy obtained from the supply would be instantaneously converted into mechanical work. The ideal curves shown in Fig. 2.7 are an important idealization. It was recognized by Byrne that such characteristics would give the switched reluctance motor some of the properties of the permanent-magnet motor. If the current is constant, the flux-*linkage* of a phase winding in such a machine has a triangular or trapezoidal waveform as the rotor rotates at constant speed, giving rise to a square-wave induced EMF. This is the same as the brushless DC motor. (See also Fig. 2.4.)

In the real switched reluctance motor the aligned inductance is not infinite and the magnetization curves are nowhere near ideal (though with cobalt-iron machines having small airgaps, this ideal is more closely approached). In design it is desirable to aim for an inductance ratio of at least 10, and this is possible only with relatively low pole-numbers and small airgaps.

The special case of the non-saturable motor.

In a motor with no magnetic saturation the magnetization curves would be straight lines as shown in Fig. 2.8. At any position the coenergy and the stored magnetic energy are equal, and are given by

$$W_f \ = \ W' \ = \ \frac{1}{2}Li^2 \qquad [8]$$

where L is the inductance at a particular position. In this case the instantaneous torque reduces to

$$T \ = \ \frac{1}{2} \, i^2 \, \frac{dL}{d\theta} \qquad [9]$$

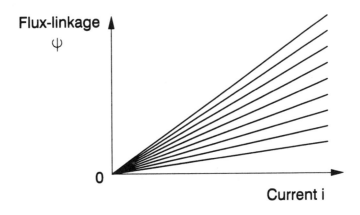

Fig. 2.8   Magnetization curves of non-saturable motor

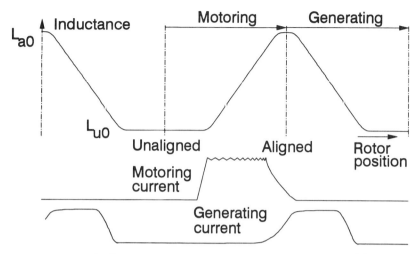

Fig. 2.9   Inductance vs. rotor position in non-saturable motor

The inductance curves $L(\theta)$ for the non-saturable motor are of the form shown in Fig. 2.9. During the period of overlap $dL/d\theta$ = constant, so the torque is constant if the current is held constant during this period. Although equation [9] is often quoted for switched reluctance motors, the non-saturating motor is not of much practical use: if a motor were made with sufficiently large airgap and thick magnetic sections to avoid saturation, its torque per unit volume would be very low and its controller would also be over-sized.

2.3 Average Torque. From §2.2 it is clear that the instantaneous torque is not necessarily constant. Setting aside the question of torque ripple, the average torque is more important from the user's point of view. Although an expression for the average torque could be derived mathematically by integrating equation [3], it is more illuminating to derive it from areas on the energy-conversion diagram (i-ψ diagram). This is done in three stages in Fig. 2.10.

Suppose that the motor is rotating at essentially constant speed and that voltage is applied to phase 1 at or near the unaligned position θ_u. The flux-linkage ψ increases according to the equation

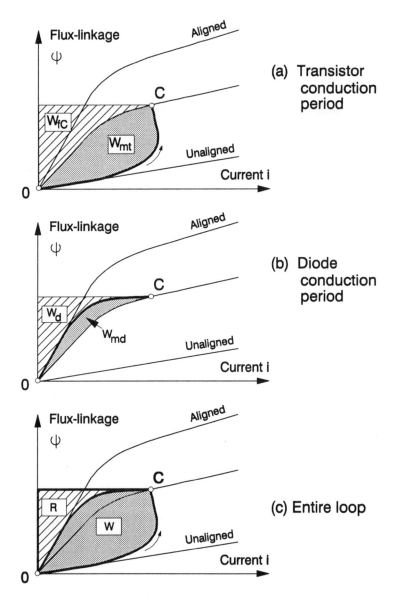

Fig. 2.10   Average torque; energy-conversion loop

$$\psi = \int (V_s - Ri) \, dt = \frac{1}{\omega} \int (V_s - Ri) \, d\theta. \qquad [10]$$

If the supply voltage V_s is constant and the phase resistance R is small, then ψ increases linearly with rotor position. The current rises linearly at first, while the inductance around the unaligned position remains low and nearly constant. But as the poles overlap, the inductance increases and a back-e.m.f. builds up, reducing the rate of rise of current. The locus of the operating point $[i,\psi]$ therefore follows the curve shown in Fig. 2.10(a) between 0 and C. At C the phase is commutated: that is, the supply voltage is reversed and the current freewheels through the diode. Fig. 2.10(a) shows the magnetization curves for the aligned and unaligned positions, together with the one for the commutation angle θ_C.

At C the accumulated energy from the supply is equal to the total area $U = W_{mt} + W_{fC}$. The stored magnetic energy is equal to W_{fC}. Therefore the mechanical work done between O and C is W_{mt}; this is during the period of transistor conduction. Note that in Fig. 2.10 this is roughly comparable to W_{fC}, meaning that only half of the energy supplied has been converted to mechanical work. The other half is stored in the magnetic field.

After commutation, Fig. 2.10(b), the supply voltage is reversed and the energy W_d is returned to the supply. At O the current and flux are both extinguished and there is no stored magnetic energy. The mechanical work done between C and O is equal to $W_{md} = W_{fC} - W_d$. Note that in Fig. 2.10(b) this is less than one-half of W_{fC}.

A simple energy balance can be deduced from the estimated areas in Figs. 2.10(a) and (b). Suppose that the energy supplied from the controller during the "fluxing" period (transistor conduction period) between O and C is $U = W_{mt} + W_{fC} = 10$ Joules. At C, 5J has been converted to mechanical work and 5J are stored in the field. During the "de-fluxing" period (diode conduction period) between C and O, $W_d = 3.5$J is returned to the supply and $W_{md} = 1.5$J is converted to mechanical work. The total mechanical work is therefore $W = W_{mt} + W_{md} = 5 + 1.5 = 6.5$J or 65% of the energy originally supplied by the controller. The energy returned to the supply is $W_d = 3.5$J or 35% on each "stroke".

The entire stroke is shown in Fig. 2.10(c), which combines the two earlier diagrams. The energy conversion is now shown as the area W, while the energy returned to the supply is $R = W_d$. The original energy supplied by the controller is $U = W+R$. Lawrenson proposed the term *energy ratio* E to characterize the "availability" of this original energy for conversion:

$$E = \frac{W}{W + R} = \frac{W}{U}. \qquad [11]$$

The energy ratio is in some sense analogous to power factor in AC machines, but it is in fact a more general concept and can be used to analyse the energy flow in AC machines also. In the simple accounting done earlier, $E = 0.65$.

The average torque can now be determined from the number of energy-conversion loops per revolution, that is, the number of strokes per revolution. In one revolution all N_r poles must be worked on by all m phases, the number of strokes per revolution is mN_r, and therefore the average torque over one revolution is

$$T = \frac{mN_r}{2\pi} \cdot W. \qquad [12]$$

If W is in Joules, T is in N-m. [1Nm = 0.736 ft-lb].

This theory presupposes that the energy-conversion loops are the same for all phases. In practice there may be a degree of interaction between phases which causes this not to be so. The interaction is possible even in an ideal machine which has no manufacturing imperfections such as run-out, and it depends on the numbers of poles and phases and how the coils are wound and connected. It is generally desirable to minimize the interaction between phases by providing sufficient cross-section in the shared parts of the magnetic circuit, and by minimizing eccentricity or geometrical asymmetries in manufacturing.

2.4 Controller volt-amperes. The original energy U supplied by the converter can be expressed as a fraction k of the product $i_C\psi_C$, where i_C and ψ_C are the values of current and flux-linkage at commutation. If the flux-linkage increases linearly during the "fluxing" period between O and C in Fig. 2.10, then $\psi_C = V_s\delta/\omega$ where δ is the conduction angle ("dwell") of the transistors during the "fluxing" period and ω is the angular velocity. Therefore

$$U = \frac{W}{E} = \frac{kV_s\delta i_C}{\omega} \qquad [13]$$

and since i_C is the peak current from the controller, the peak volt-ampere rating of the m-phase controller is indicated by

$$Q_m = mV_s i_C = \frac{mW\omega}{Ek\delta} = \frac{2\pi T\omega}{N_r Ek\delta}. \qquad [14]$$

The product $T\omega$ is the airgap power P_{gap}, while the produce $N_r\delta$ is constant with a maximum value of approximately $\pi/2$ at the "base speed"[1]; therefore the controller peak volt-amperes Q_m is given by

$$Q_m = \frac{4P_{gap}}{kE}. \qquad [15]$$

Q_m appears to be independent of the number of poles and phases, and inversely proportional to the energy ratio E and the "utilization ratio" k. Both E and k depend critically on the static magnetization curves and, in particular, on the unaligned and aligned curves. It is found that the magnetization curves are by no means independent of N_r, which therefore has a significant, if indirect, influence on the controller volt-amperes. As the number of poles increases, the inductance ratio λ decreases, and both k and E decrease.

The peak kVA/kW Q_m/P_{gap} can be estimated assuming $k = 0.7$ and $E = 0.6$: the result is 9.5 kVA/kW. If 95% of P_{gap} reaches the shaft, then the overall kVA/kW figure is 10. This value is typical of actual switched reluctance motors, and *a similar value is typical of AC induction motors operating at their base speed.*

A more detailed analysis along these lines can be found in recent works by Miller [16] and by Harris et al. [17,18].

---

[1] Base speed is defined to be the maximum speed at which rated torque can be achieved, without exceeding rated voltage and rated current. At this condition the transistor and diode conduction angles are approximately equal, and have a maximum value of one-half the rotor pole-pitch.

3 Switched Reluctance Motor Design

3.1 Definitions. A **regular** switched reluctance motor is one in which the rotor and stator poles are symmetrical about their centre-lines and equally spaced around the rotor and stator respectively. An **irregular** motor is one which is not regular. Examples of both types will be seen in this chapter.

The **absolute torque zone** is the angle through which one phase can produce non-zero torque. In a regular motor the maximum torque zone is π/N_r. The **effective torque zone** is the angle through which one phase can produce *useful* torque comparable to the rated torque. The effective torque zone is comparable to the lesser pole-arc of two overlapping poles. For example, in Fig. 2.1 the effective torque zone is equal to the stator pole-arc $\beta_s = 30°$.

The **stroke angle** ε is given by $2\pi/(\text{strokes/rev})$ or

$$\varepsilon = \frac{2\pi}{m N_r}. \qquad [1]$$

The **absolute overlap ratio** ρ_A is defined as the ratio of the absolute torque zone to the stroke angle: evidently this is equal to $m/2$. A value of at least 1 is necessary if the *regular* motor is to be capable of producing torque at all rotor positions. In practice a value of 1 is not sufficient, because one phase can never provide rated torque throughout the absolute torque zone. The **effective overlap ratio** ρ_E is defined as the ratio of the effective torque zone to the stroke angle. For regular motors with $\beta_s < \beta_r$ this is approximately equal to β_s/ε. For example, in Fig. 2.1 the effective overlap ratio is $30°/30° = 1$. Note that $\rho_E < \rho_A$. A value of ρ_E of at least 1 is necessary to achieve good starting torque from all rotor positions with only one phase conducting, and it is also a necessary (but not sufficient) condition for avoiding torque dips.

3.2 Numbers of phases and poles.

3.2.1 Two-phase and single-phase motors

With $m = 2$, $\rho_A = 1$ and $\rho_E < 1$, and therefore the two-phase regular motor is impractical unless it is provided with a parking mechanism or some form of starting assist. A primitive 2-phase motor is shown in Fig. 3.1. Phase 1 comprises two coils wound on poles 1 and 3. Phase 2 is wound on poles 2 and 4. The main flux path is indicated by the heavy line. At two "singular points" the torque is zero. One of these is the position shown in Fig. 3.1: the aligned position on phase 1 is the unaligned position on phase 2 (and *vice-versa*). In theory the displacement needed to clear these points is infinitesimal, but in practice the torque is small for several degrees on either side of them; see Fig. 3.4.

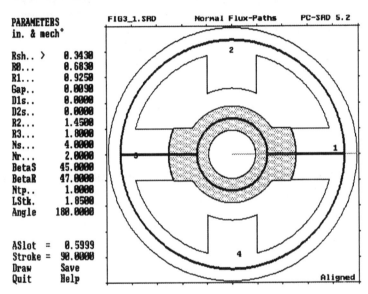

Fig. 3.1   2-phase 4/2 motor; heavy line shows the flux path in the aligned position

The 2-phase motor is desirable because of the savings in connections and transistors, and Fig. 3.2 shows an *irregular* 4/2 motor with a stepped airgap. The aligned inductance is slightly increased compared with that of Fig. 3.1, but the unaligned inductance is also greater, so there may be no gain in inductance ratio compared with Fig. 3.1. However, the "dead zone" near the unaligned position is reduced, as shown in Fig. 3.4. The stepped airgap is probably an old idea but significant work on it has been reported and the description by El-Khazendar and Stephenson is quoted here (see [20]):

Page 26

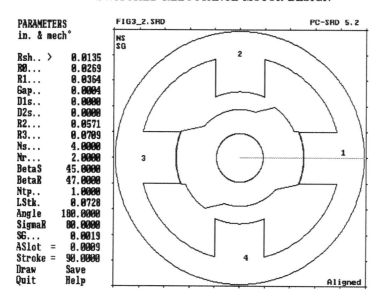

PARAMETERS in. & mech°	
Rsh.. >	0.0135
R0...	0.0269
R1...	0.0364
Gap..	0.0004
D1s..	0.0000
D2s..	0.0000
R2...	0.0571
R3...	0.0709
Ns...	4.0000
Nr...	2.0000
BetaS	45.0000
BetaR	47.0000
Ntp..	1.0000
LStk.	0.0728
Angle	180.0000
SigmaR	80.0000
SG...	0.0019
ASlot =	0.0009
Stroke =	90.0000
Draw	Save
Quit	Help

FIG3_2.SRD   PC-SRD 5.2

NS
SG

Aligned

Fig. 3.2   2-phase 4/2 motor with stepped gap

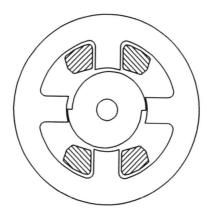

Fig. 3.3   2-phase snail-cam motor. Only one phase winding is shown.

Page 27

"The effect of stepping the airgap is to extend the region of positive inductance variation such that at any rotor position there is a positive dL/dθ for either phase winding."

It is also possible to use a uniformly tapered gap like a snail-cam, as shown in the 4/2 motor, Fig. 3.3. Bedford described a 3-phase variant of this machine [6]. For rotation in one direction, each phase can produce unidirectional torque over an angle greater than $\pi/2$, making $\rho_E > 1$ *in one direction only.* Continuous torque is available even at zero speed.

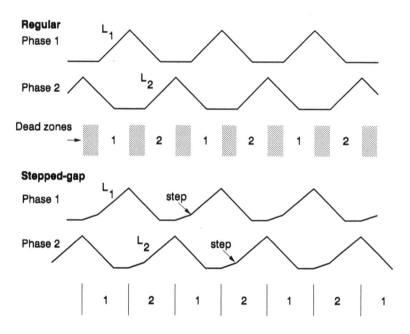

Fig. 3.4   Stepped gap eliminates dead zones

Professor Byrne described two other forms of *irregular* 4/2 motor, Figs. 3.5 and 3.6. In both cases the effect is similar to that of the stepped airgap, but there is a significant difference. Byrne and his co-workers were attempting to obtain a region of magnetic saturation at the overlapping stator and rotor pole-corners, so that the

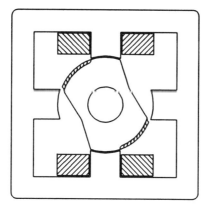

Fig. 3.5   Two-phase motor described by Prof. Byrne [9,19] with tapered pole-arc

magnetization curves would be "ideal" in the sense of Fig. 2.7. This helps to keep the torque/ampere more even as the rotor rotates, and helps minimize the volt-ampere rating of the controller. Stepped-gap motors have no saturation while the leading edge of the rotor is under the stator pole, because of the length of the airgap.

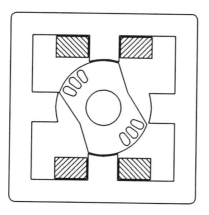

Fig. 3.6   2-phase motor with controlled saturation

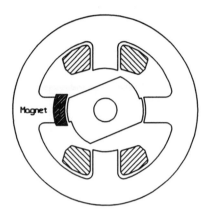

Fig. 3.7   1-phase motor with parking magnet [22]

The large slot areas of the 2-phase motor can be exploited to get low copper losses. The large iron sections keep the core losses low and maintain good mechanical stiffness, which is important for minimizing acoustic noise. The core losses are also minimized by having a relatively low commutation frequency, and the inductance ratio is high because of the large airgaps in the unaligned position. Although self-starting two-phase machines are directional, it is unclear why they appear to have received little recent attention.

It might be expected that the lowest-cost motor and controller would be **single-phase**. Only one transistor and one diode would be needed in the controller, and the number of coils and connections would be a minimum.

As in other types of motor, a true **single-phase** switched reluctance motor is possible only if the starting problem can be overcome *and* there is sufficient load inertia to coast the motor through the "blank zones" between successive torque zones. These blank zones are inevitable in single-phase motors, because $\rho_A < 1$. There is no possibility of producing constant torque throughout one revolution.

Furthermore, starting is clearly impossible without some special "assist". Compter [21] produced a single-phase motor in which the rotor is aligned with a parking magnet when stationary. When the controller is switched off, the rotor coasts to a low speed and is captured by the parking magnet in a position from which it can restart. The parking position is in the *effective torque zone* of the single phase winding.

A more recent single-phase motor patented by Horst [22] is shown in Fig. 3.7. The orientation is such that when power is first supplied to the phase winding, a sufficient impulse is imparted so that the rotor continues to rotate through the blank zone until the torque zone is re-entered, and another pulse of current then provides another accelerating impulse. In Horst's motor, the soft-iron laminated pole opposite the parking magnet is used to carry a Hall sensor or search coil as position sensor for commutation.

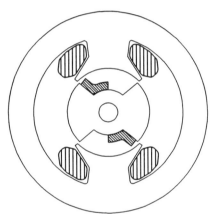

Fig. 3.8   1-phase motor with eddy-current reaction starting assist

A second method for starting a single-phase motor is shown in Fig. 3.8. In this case the rotor has conductive inserts of aluminium or copper in the spaces between the rotor poles. These inserts are electrically asymmetrical. When voltage is applied to the phase winding, the eddy-currents induced in the inserts produce a reaction torque against the stator poles. Because the eddy-currents decay, there is no position at which a stable equilibrium can be maintained between the eddy-current reaction torque and the normal magnetostatic torque. The result is that a starting impulse can be obtained at any rotor position. The disadvantage of this technique is the need for the conducting inserts and the additional losses induced in them under normal running.

Many other forms of single-phase switched reluctance motor have been devised, including limited-rotation actuators and ratchet-motors.

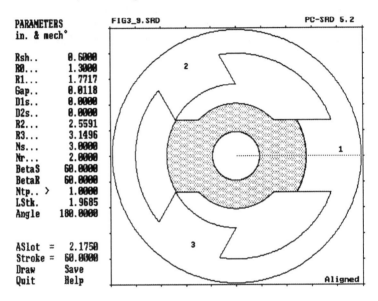

PARAMETERS		FIG3_9.SRD	PC-SRD 5.2

in. & mech°

Rsh.. 8.6888
R0... 1.3888
R1... 1.7717
Gap.. 8.8118
D1s.. 8.8888
D2s.. 8.8888
R2... 2.5591
R3... 3.1496
Ns... 3.8888
Nr... 2.8888
BetaS 68.8888
BetaR 68.8888
Ntp.. > 1.8888
LStk. 1.9685
Angle 188.8888

ASlot = 2.1758
Stroke = 68.8888
Draw Save
Quit Help

Aligned

Fig. 3.9　　Primitive 3-phase 3/2 motor

3.2.2 Three-phase motors

With $m = 3$, $\rho_A = 1.5$ and ρ_E can have values of 1 or more, so *regular* three-phase motors can be made for 4-quadrant operation.[1] The discussion considers motors in order of the rotor pole-number.

---

[1] *4-quadrant* means operation with positive and negative torque, and with positive or negative speed; that is, motoring or generating in either the forward or reverse direction.

A primitive 3-phase motor is shown in Fig. 3.9, with only three stator poles and two rotor poles, giving 3 strokes/rev. Unbalanced magnetic side-pull makes this motor impractical except for light duty. To balance the radial forces produced by each phase the stator poles are doubled to give the 6/2 motor in Fig. 3.10. This motor uses a stepped airgap to reduce the "dead zone" around the unaligned position. It has 6 strokes/rev and the rotation is in the same direction as the phase sequence: if the phases are energized in the sequence 1,2,3 the rotor rotates CCW (positive direction). The stepped gap extends the effective torque zone to approximately 55°, (ρ_E = 0.92). Alternatively both the rotor and the stator poles of Fig. 3.9 can be doubled to give the 6/4 motor in Fig. 3.11, a common arrangement. Forward rotation now corresponds to negative phase sequence. This is characteristic of **vernier** motors, in which the rotor pole-pitch is less than π/m.

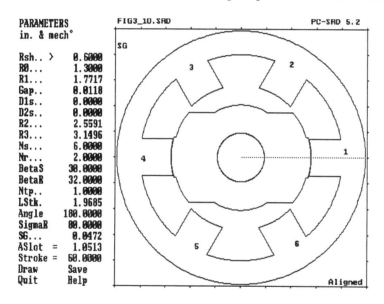

PARAMETERS in. & mech°	
Rsh.. >	0.6000
R0...	1.3000
R1...	1.7717
Gap..	0.0118
D1s..	0.0000
D2s..	0.0000
R2...	2.5591
R3...	3.1496
Ns...	6.0000
Nr...	2.0000
BetaS	30.0000
BetaR	32.0000
Ntp..	1.0000
LStk.	1.9685
Angle	180.0000
SigmaR	80.0000
SG...	0.0472
ASlot =	1.0513
Stroke =	60.0000
Draw	Save
Quit	Help

FIG3_10.SRD   PC-SRD 5.2

Fig. 3.10   3-phase 6/2 motor. Note the stepped gap

The 3-phase 6/4 motor has mN_r = 12 strokes/rev, with a stroke angle of 30°, giving $\rho_E = \beta_s/\varepsilon$ = 30/30 = 1.

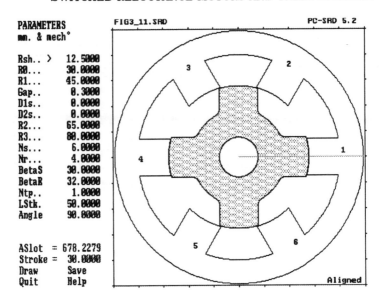

Fig. 3.11   3-phase 6/4 motor

With 6 stator poles it does not make sense to have 6 rotor poles: such a motor could have only one phase with, $\varepsilon = 60°$. With $\beta_s = 30°$, $\rho_A = 0.5$, indicating large regions of zero torque.[2] Therefore the next motor in the series with $N_s = 6$ is the 6/8 motor shown in Fig. 3.12; it has 24 strokes/rev and is similar to Konecny's motor used in the Hewlett-Packard Draftmaster plotter [13,14]. With regular motors there is always the choice of having either $N_r = N_s$ - 2, as in the 6/4; or $N_r = N_s$ + 2, as in the 6/8. The advantage of the larger N_r is a smaller stroke angle, leading possibly to a lower torque ripple; but inevitably the price paid is a lower inductance ratio which may increase the controller volt-amperes and decrease the specific output. The stator pole arc has to be reduced below that of the 6/4 motor and this decreases the aligned inductance and the inductance ratio (although it increases the slot area). The consequent reduction in

---

[2] Motors with $N_s = N_r$ can be used in limited-rotation actuators. They can also be used if they have multiple stacks with rotors offset by π/m.

available conversion energy tends to cancel the increase in the number of strokes/rev, and the core losses may be higher than those of the 6/4 or 6/2 motor because of the higher switching frequency.

Fig. 3.12   3-phase 6/8 motor

Next in the series of 3-phase motors with 6-pole stators, the 6/12 is not practical because all phases are in the aligned position at once, as in the 6/6. The 6/14 is a feasible concept, but with such a large number of rotor poles the stator pole becomes narrow and flimsy, and the inductance ratio falls to a low value. Because of the progressive reduction of inductance ratio and energy-conversion per stroke, it is not worth considering rotor pole numbers above 8 in this series.

The 12/8 three-phase motor is effectively a 6/4 with a "multiplicity" of two. It has 24 strokes/rev, with a stroke angle of 15°, and $\rho_A = 1.5$. In Fig. 3.13, $\rho_E = 15/15 = 1$, the same as for the 6/4 motor discussed earlier. A high inductance ratio can be maintained and the end-windings are short: this minimizes the copper losses, shortens the frame, and decreases the unaligned inductance. Moreover, the magnetic field in this machine has *short flux paths* because of its *four-pole* magnetic field configuration, unlike the two-pole

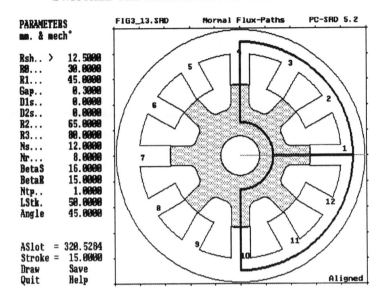

Fig. 3.13   3-phase 12/8 motor. The heavy line represents two of the four flux-paths.

configuration in the 6/4 (or the 8/6; see below). Although the MMF per pole is reduced along with the slot area, the effects of long flux-paths through the stator yoke are alleviated.

Fig. 3.14 shows a 12/10 motor with two teeth per stator pole, as developed and analyzed by Harris and Finch [24]. It has 30 strokes/rev, and with $\beta_s = 12°$, $\rho_E = 13/12 = 1.08$. The principle is that for a given stator electric loading (MMF), bifurcating the stator poles doubles $dL/d\theta$ and therefore the torque (Eq [9], Ch.2). However, this increase is not realised in practice because the multiple-tooth structure, with its fine geometry, leads to an increase in the unaligned inductance and a decrease in the inductance ratio, while the increase in strokes/rev increases the core losses. Furthermore, the shape of the stator poles leads to a loss of winding area, which increases the copper losses. Although there may be some gain in torque per ampere, the benefits are liable to be restricted to low speeds.

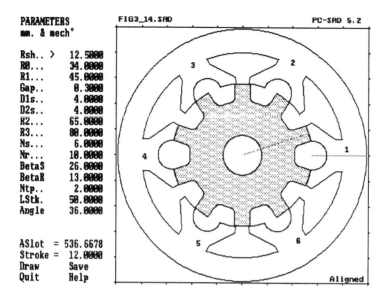

PARAMETERS   FIG3_14.SRD   PC-SRD 5.2
mm. & mech°

Rsh.. >	12.5000
R0...	34.0000
R1...	45.0000
Gap..	0.3000
D1s..	4.0000
D2s..	4.0000
R2...	65.0000
R3...	80.0000
Ns...	6.0000
Nr...	10.0000
BetaS	26.0000
BetaR	13.0000
Ntp..	2.0000
LStk.	50.0000
Angle	36.0000

ASlot = 536.6678
Stroke = 12.0000
Draw   Save
Quit   Help

Fig. 3.14   3-phase 12/10 motor with 2 teeth per stator pole

Another motor with *short flux paths* is shown in Fig. 3.15 [25]. This 3-phase 12/10 motor has its stator poles grouped in pairs and wound on two opposite pairs of stator poles, such that the flux circulates in two independent loops. Each flux loop goes through adjacent rotor and stator teeth. A more detailed view is shown in Fig. 3.16.

This is one example of a whole series of machines with different pole-number combinations [25]. Another example is the *exterior-rotor* 3-phase motor shown in Fig. 3.17: with N_s = 24 and N_r = 32, it has 96 strokes/rev. The main purpose of *short flux-paths* is to reduce the core losses and the absorption of MMF in the stator yoke. The short flux-paths are achieved by means of the winding configuration and the lamination geometry; the electronic controller is exactly the same as for any other switched reluctance motor having the same number of phases. *(Hendershot U.S. Patents 4,883,999 & 5,111,095)*

Page 37

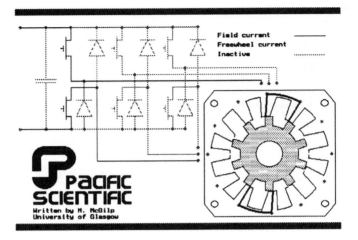

Fig. 3.15   3-phase 12/10 motor; short flux-paths

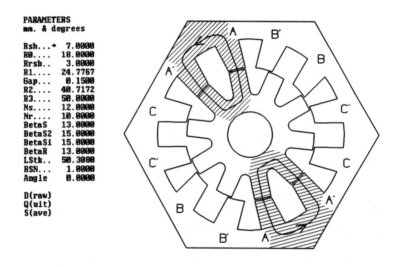

PARAMETERS
mm. & degrees

Rsh...→   7.0000
R0....   18.0000
Rrsb..   3.0000
R1....   24.7767
Gap...   0.1500
R2....   40.7172
R3....   50.0000
Ns....   12.0000
Nr....   10.0000
BetaS   13.0000
BetaS2   15.0000
BetaSi   15.0000
BetaR   13.0000
LStk..   50.3000
RSN...   1.0000
Angle   0.0000

D(raw)
Q(uit)
S(ave)

Fig. 3.16   3-phase 12/10 motor

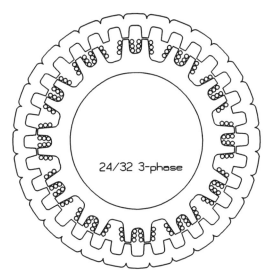

Fig. 3.17   3-phase 24/32 motor with exterior rotor

3.2.3 Four-phase motors

The 4-phase regular 8/6 motor shown in Fig. 3.18 has 24 strokes/rev and a stroke angle of 15°, giving ρ_A = 2. With β_s = 21°, ρ_E = 1.33, which is sufficient to ensure starting torque from any rotor position, and no problem with torque dips. This configuration is used in the well-known *OULTON* motor produced by Graseby Controls, England. The pole number is low enough to ensure a high inductance ratio.[3]

The next 4-phase motor is the 8/10, with 32 strokes/rev and ε = 11.25°. The inductance ratio is inevitably lower than in the 8/6: the poles are narrower, while the clearance between pole-corners in the unaligned position is smaller, increasing the unaligned inductance. This motor is probably on the borderline where these effects cancel each other out; with higher pole-numbers, the loss of inductance ratio and energy-conversion area tends to swamp the increase in strokes/rev. For this reason, higher pole-numbers are not discussed.

---

[3] With an airgap of 0.013", the first 7.5kW OULTON motors had an inductance ratio of about 12, a value which is difficult to exceed in this configuration.

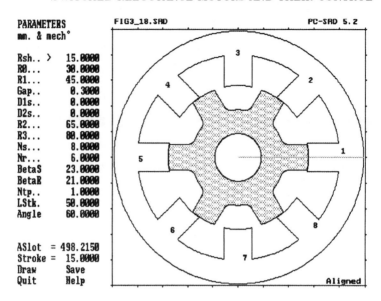

PARAMETERS
mm. & mech°

Rsh.. > 15.0000
R0... 30.0000
R1... 45.0000
Gap.. 0.3000
D1s.. 0.0000
D2s.. 0.0000
R2... 65.0000
R3... 80.0000
Ns... 8.0000
Nr... 6.0000
BetaS 23.0000
BetaR 21.0000
Ntp.. 1.0000
LStk. 50.0000
Angle 60.0000

ASlot = 498.2150
Stroke = 15.0000
Draw   Save
Quit   Help

FIG3_18.SRD   PC-SRD 5.2

Aligned

Fig. 3.18   4-phase 8/6 motor

3.2.4 Higher phase numbers

One reason for considering higher phase numbers is indicated by
Eqn. [1] on p. 24. The number of strokes per revolution can be
increased without increasing the number of poles N_r. This means
that the problem of torque dips is alleviated without the penalty of
the loss of inductance ratio that arises with high pole numbers.[4] Fig.
3.19 shows the principle in terms of the overlap between phases.

The torque waveforms in Fig. 3.19 are scaled as though constant
current was applied to each phase in turn for the whole of its
absolute torque zone (π/N_r, or 180 electrical degrees). This is not
necessarily the best way to operate the motor, but it illustrates the

---

[4]It remains the case that the minimum number of stator poles is proportional to
the number of phases, and therefore the 4-phase motor should have an advantage over
the 5-phase motor in terms of the maximum achievable inductance ratio, *ceteris
paribus*.

Page 40

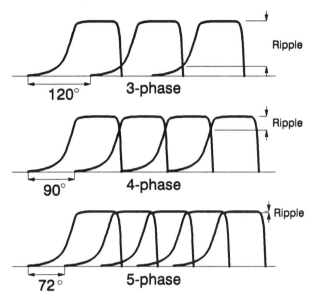

Fig. 3.19   Torque waveforms of 3,4 and 5-phase motors (after [26])

problem of torque dips in 3-phase motors if the poles are not specially widened. The waveforms for 4- and 5-phase motors show that the torque dips can be avoided and uniform torque is achievable without boosting the current in regions of low torque per ampere. Of course, there is still the question of the optimum current waveform. Constant current applied through 180 elec° would produce torque *peaks* - the opposite of torque dips - but it is better to have this problem than to be unable to start the motor from certain rotor positions.

This advantage of 4- and 5-phase motors arises not only from the pole geometry but also from their ability to operate with two phases on at the same time. In the 5-phase motor, as many as three phases may be on at the same time for short periods; see Fig. 3.19. Fig. 3.20 shows a 5-phase 10/8 motor in which the pole windings are connected to give adjacent flux-loops (heavy solid lines) which add together in the central stator pole. Each flux-loop is excited by one phase. The rotor is shown in a position where two phases are on, and both are producing positive torque in the counterclockwise direction. (A similar set of flux-loops is active on the opposite side of the machine).

Page 41

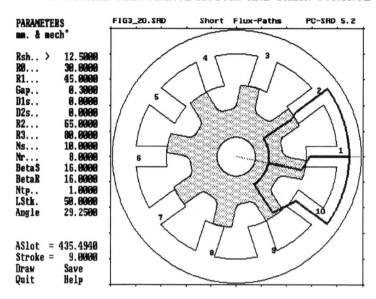

Fig. 3.20   5-phase 10/8 motor showing short flux-paths both
             producing positive torque

A further possible advantage of the 5-phase motor may be inferred
from experience with stepper motors. One of the advantages of the
5-phase stepper motor, as manufactured by Berger-Lahr, is stable
open-loop operation (with no shaft position feedback), without
resonances. The switched reluctance motor is treated in this book as
having an electronic commutator positively synchronized with shaft
position, and in that form it does not need a high phase number to
ensure stable operation. However, for *sensorless* operation there may
well be advantages in having 4 or 5 phases.

The motor shown in Fig. 3.21 is a 5-phase 10/8 motor with short flux-
paths, generally similar to the motor in Fig. 3.20. However, because
the rotor has two separate pole-pitches the number of strokes per
revolution is only 20, instead of 40 in the motor in Fig. 3.20. The core
losses should be lower by a factor of at least 2.

An even more dramatic advantage is indicated when this motor is
compared with the conventional 10/8 with normal 2-pole flux-paths,

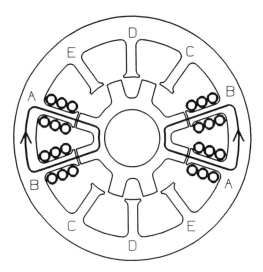

Fig. 3.21   5-phase 10.8 motor with dual rotor pole-pitch

as illustrated in Fig. 3.22. Not only is the frequency lower, but also the flux-paths are much shorter. A simple analysis illustrates this. Consider the motors in Figs. 3.20 and 3.21, and assume that both have the same maximum flux-density during operation, and the same cross-section in their flux-paths. Then the volume of iron in which core losses are generated is proportional to the total length of the flux-path in each case, and proportional to something like the square of the commutation frequency. Fig. 3.22 shows the idealised shapes of the flux-paths, and it is easy to show that for the proportions illustrated the ratio of the total flux-path lengths is

$$\frac{\frac{2\pi}{m}(R_1 + R_2) + 4(R_2 - R_1)}{2R_2(2 + \pi)} \approx 0.5 \qquad [2]$$

so that with a commutation frequency ratio of 2 the short flux-path motor has only 1/8 of the core losses at a given speed.

Michaelides and Pollock [30] reported a 4kW seven-phase 14/12 motor of the form shown in Figs. 3.23 and 3.24, in which at 500rpm the total torque could be doubled at the same r.m.s. phase current and total losses, by re-configuring the windings for short flux-paths.

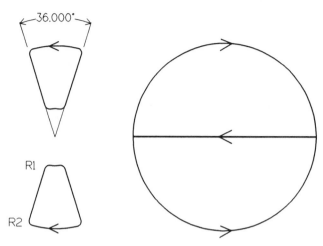

Fig. 3.22   Flux-paths of conventional 10/8 motor (right) and the
motor of Fig. 3.21 (left)

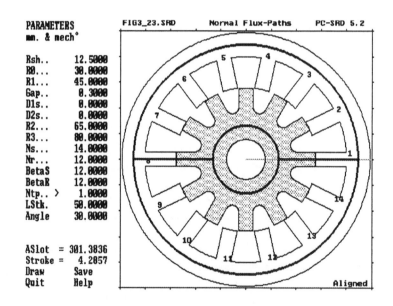

Fig. 3.23   7-phase 14/12 motor with normal 2-pole flux paths

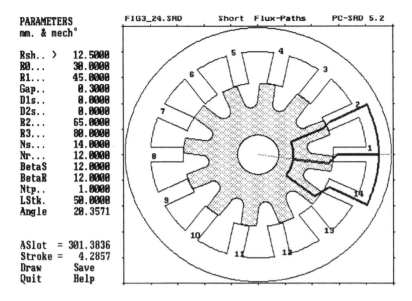

FIG3_24.SRD   Short Flux-Paths   PC-SRD 5.2

PARAMETERS
mm. & mech°

Rsh.. >	12.5000
R0...	30.0000
R1...	45.0000
Gap..	0.3000
D1s..	0.0000
D2s..	0.0000
R2...	65.0000
R3...	80.0000
Ns...	14.0000
Nr...	12.0000
BetaS	12.0000
BetaR	12.0000
Ntp..	1.0000
LStk.	50.0000
Angle	20.3571

ASlot = 301.3836
Stroke = 4.2857
Draw   Save
Quit   Help

Fig. 3.24   7-phase 14/12 motor with short flux-paths (see[30])

3.3 Pole Arcs. The rotor and stator pole arcs should be approximately the same. With reference to Fig. 3.25, consider the aligned position, **4**. This is a position of stable equilibrium. If the rotor is displaced to either side, the overlap angle changes, producing a change in inductance and a restoring torque. If the rotor pole arc was larger than the stator pole arc, $\beta_r > \beta_s$, there would be no change in inductance until the rotor had moved to a position $\beta_r - \beta_s$ away from the aligned position. In other words, there would be a dead zone centred on the aligned position. The result is similar if $\beta_s > \beta_r$. The dead zone serves no purpose other than increasing the time available for the flux to be reduced to zero after commutation, before negative torque begins to be generated. But this benefit is marginal.

Fig. 3.25 also illustrates the fact that the effective torque zone is approximately equal to the lesser of the two pole arcs. As the rotor pole moves from left to right, torque begins to be developed just before position **2**, and continues until position **4**. The angle of rotation between these positions is equal to the lesser pole arc (β_s as shown).

Page 45

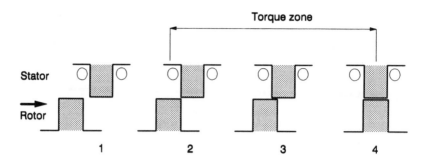

Fig. 3.25   Torque zone

The optimum pole arcs are a compromise between various conflicting requirements. On the one hand, they should be made as large as possible to maximize the aligned inductance and the maximum flux-linkage. However, if they are too wide there is not enough clearance between the rotor and stator pole-corners in the unaligned position. Motor B in Fig. 3.26 has this problem.

If the poles are narrow the slot area is increased, so the resistance and copper losses can be decreased. The rotor weight and inertia are also reduced. But with narrow poles the aligned inductance and the inductance ratio are both small. Motor C in Fig. 3.26 has this problem.

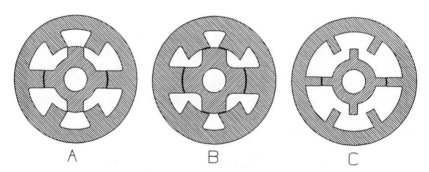

Fig. 3.26   Examples of different pole-arc combinations

The optimum pole arcs are somewhere between these extremes, as illustrated by the well-proportioned design A in Fig. 3.26. However, there is no single value that is appropriate for all applications. For very high-efficiency designs the slot area needs to be maximized and this leads to a narrower pole-arc; but the starting capability may be compromised because of torque dips, and there may be excessive torque ripple. Wider pole arcs may alleviate this problem, but at the expense of slot area and higher copper losses. The choices depend on the entire torque/speed range and on the number of poles and phases.

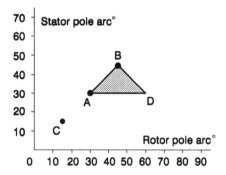

Fig. 3.27   Feasible triangle for 3-phase 6/4 motor

Lawrenson [2] gave some guidance on the choice of pole arcs in the form of "feasible triangles". Fig. 3.27 is the feasible triangle for the 3-phase 6/4 motor. It is a map of all possible combinations of pole arcs β_r and β_s. According to certain rules, the combinations must lie within the triangular boundary ABD. The diagram is drawn for $\beta_s < \beta_r$; this constraint is represented by the line AB. The effective torque zone is approximately equal to the lesser of the two pole arcs, β_s, and this must be greater than the stroke angle ε, otherwise there will be positions from which the motor cannot start. This condition is represented by the line AD, since $\varepsilon = 30°$ in the 6/4 motor. Finally, the angular separation between the corners of adjacent rotor poles must be wider than the stator pole arc, so that there is finite clearance (no overlap) in the unaligned position. This condition is represented by the equation

$$\frac{2\pi}{N_r} - \beta_r > \beta_s \qquad\qquad [3]$$

and by the line AD in Fig. 3.27.

In stepper motors where $N_r = N_s$ and $\beta_r = \beta_s$, it is possible to determine an optimum ratio between the pole arc and the tooth-pitch, based on the maximization of the aligned/unaligned inductance ratio assuming infinitely permeable steel. The finite permeability and saturation modify the optimum ratio, which is approximately 0.4. In switched reluctance motors this optimum does not apply, because in general N_r and N_s are unequal and in any case there is a wide variation in operating modes, so that the maximization of the inductance ratio is too simple a criterion. For good design, it is helpful to have CAD tools of two kinds: (1) dynamic analysis programs that can simulate the operation of the entire drive at various speeds and with various control settings; and (2) finite-element or boundary-element solvers for calculating the magnetization curves required to get accurate results in the dynamic analysis. Nevertheless, the "well-proportioned" design shown in Fig. 3.26 (A) is a good place to start, and designs that are far from the line AB in Fig. 3.27 are unlikely to perform well.

3.4 Pole geometry. Fig. 3.28 shows various pole geometries. The basic concept is that of A, but several improvements can be made to this shape. First, the corners at the bottom of the slot should always have a radius as shown in B; this stiffens the pole against *lateral* deflection. A slight taper (C) is also desirable to concentrate saturation near the airgap and minimize the MMF drop in the pole. Some designers include tangs or overhangs at the pole tips (D), to provide a register for slot wedges. Their magnetic effect is beneficial if they saturate and soften the torque impulse (jerk) that can occur at the start of overlap. By undercutting the stator pole, the slot area is slightly increased. However, this design is not appropriate if the windings are made for insertion after winding. A less common detail is the rounding of the pole corners (E); this is probably intended to soften the torque impulse at the start of overlap, but any benefits will be offset to some extent by the disadvantage of increasing the airgap, even if this is over a limited angle.

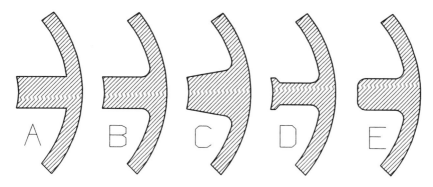

Fig. 3.28   Details of pole geometry

3.5 Windings. The windings of switched reluctance motors are simpler than those of other types of motor. Generally there is one coil on each pole. When a phase comprises two opposite poles, the two coils may be connected in series or in parallel. The maximum achievable slot fill depends on the method of winding, the type and amount of insulation, and the conductor shape.

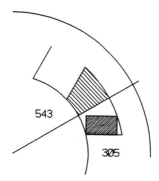

Fig. 3.29   Slot fill

In switched reluctance motors the coils can be pre-wound and then slipped over the poles without interfering with each other. This constrains the dimensions of the coil, as shown in Fig. 3.29, where the proportions are fairly typical. The rectangle (area 305 mm^2) is approximately the largest that can be inserted into the slot (half-area 543 mm^2). If the rectangle is 60% full of copper then the overall slot fill is 0.6 × 305/543 = 34%. A similar diagram for the 4-phase 8/6 motor yields a slot fill nearer 40%, and this figure is routinely achieved in production.

Usually all the leads are brought out to terminals. In this respect the switched reluctance motor is less efficient than the AC motor or the 3-phase brushless DC motor, which have an internal star point or a delta connection. Even better is the DC permanent magnet motor, which requires only two terminals. Certain controllers permit the internal connection of two or more leads; see Chapter 6.

The thermal contact between the simple rectangular coil and the stator steel is not particularly good, and J.M. Stephenson has recently patented a modification in which the unused triangular area between two coil sides is filled by an intrusion stamped in the lamination itself, Fig. 3.30. This permits the coils to be snugly surrounded by steel on three sides, with increased thermal contact. Significant increases in output are achieved by this means.

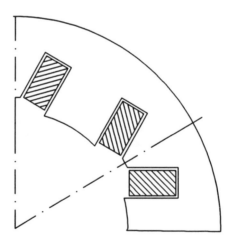

Fig. 3.30   Slot design patented by J.M. Stephenson (SRDL)

3.6 Lamination steels. The switched reluctance motor has a higher commutation frequency than AC motors of comparable speed and rating. For the same number of rotor poles, the commutation frequency in the switched reluctance motor is double that of the AC motor. Moreover, the flux-density waveform is rich in harmonics and reaches high levels of saturation especially near the pole corners. Except at high speeds, the total core losses tend to be held down by the fact that the volume of iron is less than in the comparable AC motor. But the local loss density may be high. Because of the high frequency the eddy-current component dominates, making it desirable to use a thinner lamination, preferably of Silicon steel, especially in high-efficiency applications.

For aerospace-quality designs operating at very high speeds, cobalt-iron and variants (Hiperco 50, Vanadium Permendur, Rotelloy, etc) are used. The switched reluctance motor benefits directly from the high saturation flux-density and high permeability of these alloys, because the energy-conversion diagram (Fig. 2.10) gains a significant increase in available conversion energy (area).

Because the flux is non-sinusoidal, it is not correct to use the conventional formulas for representing core losses, such as Stenimetz's equation, which is based on sinewave measurements. An adaptation of Steinmetz's equation for non-sinusoidal flux waveforms is described in [31] (see also Chapter 8), together with a procedure for extracting the necessary coefficients from sinewave loss curves at different frequencies and flux-densities. However, more work is needed before the core losses can be predicted accurately, especially at extreme levels of flux-density and at very high speeds.

3.2.5 Classification Table. The following table shows a selection of feasible pole-numbers for motors with 1-7 phases. Several other combinations are possible, but the table covers many of the combinations likely to be met in practice.[5] See also [32].

m	N_s	N_r	μ	ε °	Strokes/rev	F	Examples/comment/[Ref]
1	2	2	1	180	2	*	Horst [22]; Compter [21]
2	4	2	1	90	4	NS	Byrne [9,19]
3	6	2	1	69	6	SG	$\rho_E < 1$
3	6	4	1	30	12	OK	Uneven torque
3	6	8	1	15	24	OK	HP Draftmaster/Warner [13]
3	12	8	2	15	24	OK	Allenwest MotionMaster
3	18	12	3	10	36	OK	Low λ
3	24	16	4	7.5	48	OK	Low λ
4	8	6	1	15	24	OK	OULTON
4	16	12	2	7.5	48	OK	Low λ
5	10	4	1	20	18	OK	
5	10	6	1	12	30	OK	
5	10	8	1	9	40	OK	
5	10	8	2	18	20	OK	Magna Physics [26]
6	12	10	1	6	10	?	
6	24	20	2	3	120	?	
6	12	14	1	4.29	84	?	
7	14	10	1	5.14	70	?	
7	14	12	1	4.29	84		See Pollock [30]

m = No. of phases; N_r = rotor poles; N_s = stator poles; ε = stroke; F = Feasibility; NS = Non-symmetric rotor; SG = stepped-gap rotor. μ = No. of working pole-pairs/phase (or "multiplicity"); ε = stroke angle. * Needs "assist" for starting.

---

[5] See PC-SRD 5 manual for an extended list.

4. Dynamic Operation

4.1 Single-pulse operation. The flux in the switched reluctance motor is not constant but must be established from zero every stroke. In motoring operation the build-up is timed to coincide with the period when the rotor poles are approaching the stator poles of the phase that is due to be excited. The process is controlled by switching the supply voltage on at the *turn-on angle* θ_0 and switching off at the commutation angle θ_C.

Assuming that each phase is supplied by a circuit of the form of Fig. 4.1, both transistors are switched on at θ_0 and both are switched off at θ_C. At a sufficiently high speed, the waveforms of voltage, flux-linkage, current, and idealised inductance are shown in Fig. 4.2: operation of the motor with this set of waveforms is called "single-pulse" operation. The "idealised" inductance is the inductance that would be obtained with no fringing and with infinitely permeable stator and rotor. It is used purely as a convenient means for relating the waveforms to the rotor position, because it directly shows the amount of overlap between the stator and rotor poles on the phase in question. The aligned and unaligned positions are shown.

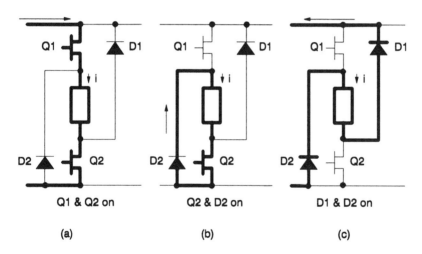

Fig. 4.1   Conduction modes in one phase

The build-up of flux-linkage proceeds according to Faraday's Law: at constant angular velocity ω this has the form

$$\psi_C = \int_{\theta_0}^{\theta_C} (V_s - Ri) \, d\theta/\omega + \psi_0 \qquad [1]$$

where ψ_0 is the flux-linkage pre-existing at θ_0 (ordinarily zero). V_s is the supply voltage, R the phase resistance, and i the instantaneous current. All impedances and volt-drops in the controller and the supply are ignored at this stage. Note that Faraday's Law describes the flux-*linkage*, not the flux. Equation [1] can be written as

$$\omega\psi_C = (V_s - v_1) \cdot \theta_D \qquad [2]$$

where $\theta_D = (\theta_C - \theta_0)$ is the *dwell* and v_1 is the mean volt-drop due to resistance during θ_D. If $Ri \ll V_s$ the flux-linkage rises linearly.

In motoring operation the flux should ideally be reduced to zero before the poles are separating, otherwise the torque changes sign and becomes a braking torque. To accomplish this the terminal voltage must be reversed at θ_C, and this is usually done by the action of the freewheeling diodes when the transistors turn off. The angle taken for the negative voltage to drive the flux back to zero at the "extinction angle" θ_q is again governed by Faraday's Law:

$$0 = \psi_C + \int_{\theta_C}^{\theta_q} (-V_s - Ri) \, d\theta/\omega \qquad [3]$$

and this can be written as

$$\omega\psi_C = (V_s + v_2) \cdot (\theta_q - \theta_C). \qquad [4]$$

where v_2 is the mean volt-drop due to resistance in the *de-fluxing period* $(\theta_q - \theta_C)$. If $Ri \ll V_s$ the flux-linkage falls linearly, and at constant speed the angle traversed is nearly equal to the dwell angle, both being equal to ψ_C/V_s. The total conduction angle covers the entire cycle of building and extinguishing the flux and is equal to

$$\theta_q - \theta_0 \approx \frac{2\omega\psi_C}{V_s}. \qquad [5]$$

The peak flux-linkage ψ_C occurs at the commutation angle θ_C.

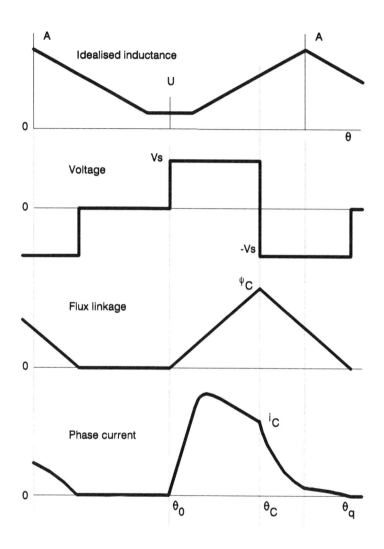

Fig. 4.2   Single-pulse waveforms

The entire conduction period must be completed within one rotor pole-pitch $\alpha_P,$[1]

$$\alpha_P = \frac{2\pi}{N_r};$$ [6]

otherwise there will be a ratcheting effect in which ψ_0 has a series of non-zero values increasing from stroke to stroke. This condition is also called "continuous conduction". That is,

$$\theta_q - \theta_0 < \alpha_P.$$ [7]

Equations [2-7] combine to give the maximum permissible dwell angle

$$\theta_D < \frac{\alpha_P}{1 + \xi}$$ [8]

where

$$\xi = \frac{1}{1 + \dfrac{V_s - v_1}{V_s + v_2}}.$$ [9]

If the mean resistive volt-drops v_1 and v_2 are both approximately the same fraction ρ of V_s, then equations [7] and [8] reduce to

$$\theta_D < \alpha_P \cdot \frac{(1 + \rho)}{2}.$$ [10]

For example, in a symmetrical 6/4 motor the pole-pitch is $\alpha_P = 90°$ (360 elec°) and if $\rho = 0$ the maximum dwell angle is $\theta_D = 45°$, giving a total angle of conduction in the phase winding of 90°. But if $\rho = 0.2$ the maximum dwell angle is 54° and the conduction angle is 108°.

In a regular switched reluctance motor the angle of rising inductance is only $\alpha_P/2$. Ideally the flux should be zero throughout the period of falling inductance, because current flowing in that period produces a negative or braking torque. To avoid this completely, the conduction angle must be restricted to $\alpha_P/2$ and the maximum dwell angle is then

---

[1]Note that α_P is the angle of rotation between two successive aligned positions.

$$\theta_D \; < \; \frac{\alpha_P}{2} \cdot \frac{1 + \rho}{2}. \qquad\qquad [11]$$

In the 6/4 motor, with $\rho = 0.2$ this indicates a maximum dwell angle of 27° (108 elec°) and a conduction angle of 54°. In practice, larger dwell angles than this are used because the gain in torque-impulse during the rising-inductance period exceeds the small braking-torque impulse, which generally occurs in a region when the torque/ampere is low (i.e., near the aligned and/or unaligned positions). This condition is shown in Fig. 4.2, where the current has a "tail" extending beyond the aligned position. The torque is negative during this tail period, but it is small.

The turn-on angle in Fig. 4.2 is at the unaligned position, and the current rises linearly until the poles begin to overlap. The rising inductance generates a back-EMF which consumes an increasing proportion of the supply voltage, until at the peak of the current waveform the back-EMF equals V_s. Subsequently the back-EMF grows greater than V_s because the flux-linkage is still increasing, while the speed is constant. What was an excess of applied forward voltage now becomes a deficit, and the current begins to decrease. At the point of commutation the applied terminal voltage reverses, and there is a sharp increase in the rate of change of current.

At the aligned position the back-EMF reverses, so that instead of augmenting the negative applied terminal voltage, it diminishes it, and the rate of fall of current decreases. In this period there is a danger that the back-EMF may exceed the supply voltage and cause the current to start increasing again. It is for this reason that in single-pulse operation, commutation must precede the aligned position by several degrees. The commutation angle must be advanced as the speed increases.

Fig. 4.2 also shows the importance of switching the supply voltage on before the poles begin to overlap. This permits the current to grow to an adequate level while the inductance is still low. For as long as the inductance remains nearly constant, there is no back-EMF and the full supply voltage is available to force the increase in current. The turn-on angle may be advanced well ahead of the unaligned position at high speed, even into the previous zone of falling inductance .

2. Chopping - Voltage PWM.[2] Chopping is necessary to control the current at low speed. The simplest scheme is to leave one transistor (Q2 in Fig. 4.1) on during the period θ_0 to θ_C, and to switch the other one (Q1) on and off at a high frequency with a fixed duty-cycle $d = t_{ON}/T$, where t_{ON} is the on-time and T is the period of the chopping frequency. This strategy is called *soft chopping*.

When Q1 is on, voltage V_s is connected to the phase winding. When it is off, the winding is short-circuited through Q2 and D2. Q1 is called the "chopping transistor" and D2 the "chopping diode". Q2 is called the "commutating transistor" and D1 the "commutating diode", because they change state only at the commutation angles θ_0 and θ_C.

The waveforms during soft chopping are shown in Fig. 4.3. During the dwell angle the average voltage applied to the phase winding is dV_s. Again using the parameter ρ to represent the averaged effect of resistive volt-drops, the flux-linkage rise in the dwell period can be equated to the flux-linkage fall in the de-fluxing period to give

$$\omega\psi_C = \theta_D(d - \rho)V_s = (\theta_q - \theta_C)(1 + \rho)V_s. \qquad [12]$$

This can be rearranged to show that the total conduction angle is

$$\theta_q - \theta_0 = \theta_D\left[\frac{1 + d}{1 + \rho}\right]. \qquad [13]$$

To prevent continuous conduction, θ_D must be restricted to

$$\theta_D < \alpha_P \cdot \frac{1 + \rho}{1 + d}. \qquad [14]$$

For example, in the 6/4 motor, if $\rho = 0.2$ and $d = 0.5$, the maximum dwell is 72°. To prevent *any* braking torque, θ_D must be restricted to

$$\theta_D < \frac{\alpha_P}{2} \cdot \frac{1 + \rho}{1 + d}, \qquad [15]$$

i.e. one-half of the absolute maximum or 36° in the example. The dwell can be increased as the duty-cycle is decreased, up to the maximum given by equation [13].

---

[2]PWM = pulse-width modulation. Note that the simple equations developed in this section make use of the principle of *state-space averaging*.

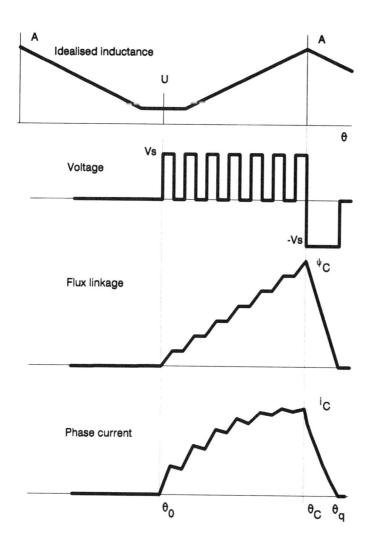

Fig. 4.3   Voltage PWM - soft chopping

A similar analysis can be carried out for *hard chopping*, Fig. 4.4, in which *both* transistors are switched together at high frequency. If the switching frequency remains the same, hard chopping increases the current ripple by a large factor (typically 5-10 times), and for this reason it is not desirable as a control strategy for motoring operation. However, in generating or braking operation it may be necessary as the only feasible means to control the current. Some controllers (notably the *OULTON*-type controller, Chapter 6) are not capable of soft chopping and operate in hard chopping all the time. It appears that soft chopping produces lower acoustic noise and less EMI. It also decreases the DC ripple current in the supply and substantially reduces the requirement for filter capacitance.

In both soft and hard chopping, the flux-linkage waveform increases in regular steps with a more-or-less constant average slope, as reflected in the left half of equation [12]. Before the start of overlap, the average slope of the *current* waveform is also nearly constant as the linearly-increasing flux is forced into a constant inductance. Thereafter, the inductance increases more or less linearly while the flux-linkage continues to rise linearly. Consequently the current tends to become constant. Unlike the single-pulse waveform or the current-regulated waveform below, the current waveform under voltage-PWM does not "hit" the overlapping pole-corners with a high current, and there is some evidence that it produces quieter operation. Not only is the current lower at the start of overlap, but also there are repeated short freewheeling periods which may help to soften the mechanical impact.

The waveforms in Fig. 4.3 and 4.4 show the turn-on angle at the unaligned position and the commutation angle at the aligned position. This illustrates the fact that at low speed, when chopping is the preferred control strategy, the whole of the absolute torque zone can be used. As is evident from equation [12], the ratio of the slopes of the rising and falling parts of the flux-linkage waveform is approximately equal to d, so that with a low duty-cycle (needed to "throttle" the voltage at low speed), the de-fluxing is accomplished in a very few degrees, permitting late commutation.

Although the pole *arcs* do not appear in any of the equations constraining the limiting values of the firing angles, they are extremely important in determining their *optimum* values.

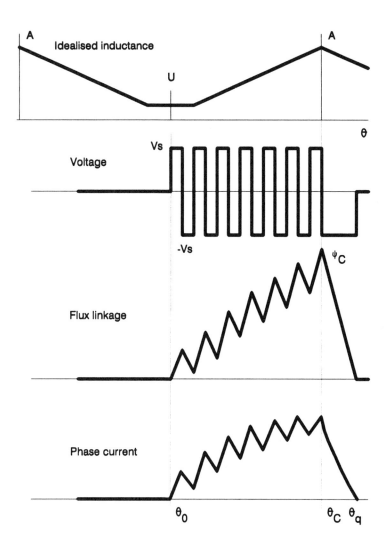

Fig. 4.4   Voltage PWM - hard chopping

4.3 Chopping - current regulation. Fig. 4.5 shows the waveforms obtained with a hysteresis-type current regulator in which the power transistors are switched off or on according as the current is greater or less than a reference current. The instantaneous phase current is measured using a wide-bandwidth current transducer, and fed back to a summing junction. The error is used directly to control the states of the power transistors.

Both soft and hard chopping schemes are possible, but only the soft-chopping waveforms are shown in Fig. 4.5. The waveforms for hard chopping are similar. As in the case of voltage-PWM, soft chopping decreases the current ripple and the filter requirements, but it may be necessary in braking or generating modes of operation.

The simple hysteresis controller maintains the current waveform between an upper and a lower limit - in the *hysteresis band*. As the supply voltage is fixed, the result is that the switching frequency decreases as the incremental inductance of the phase winding increases. (The incremental inductance increases along with the total inductance.) This can be seen in Fig. 4.5.

In Fig. 4.5 the flux-linkage waveform shows a period of constant flux-linkage before the start of overlap, as constant current is being forced into a constant inductance. Thereafter, the inductance increases while the average current remains constant, so the flux-linkage rises along with the inductance.

4.4 Current waveform. As the supply is a voltage source (albeit switched between positive, negative or zero values), current is drawn according to the effective resistance, back-EMF, and inductance of the phase winding, all of which vary as the rotor rotates. The back-EMF and inductance also vary with current because of magnetic saturation.

The flux-linkage waveform is forced to be the integral of $(V - Ri)$. Once the flux-linkage is known the current can be determined from the magnetization curves (Fig. 2.2), which are expressed as

$$\psi = \psi(i, \theta) \qquad [16]$$

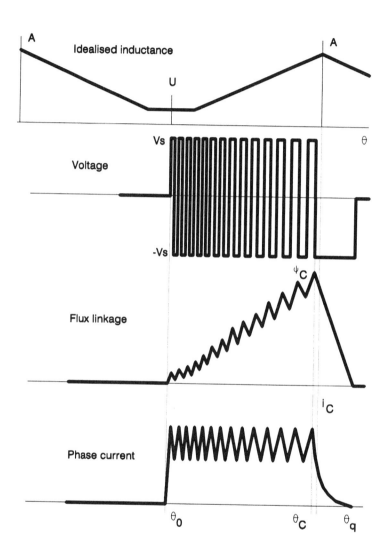

Fig. 4.5   Current regulation - soft chopping

Extracting the current from these curves requires a nonlinear algebraic solution of equation [16] when ψ and θ are known. The order of "dependence" is : voltage $\rightarrow$ flux-linkage $\rightarrow$ current. Torque and losses may then be calculated *a posteriori* from the current and flux-linkage waveforms.[3] (The detailed calculation of core losses requires that the flux-*density* waveform in all parts of the machine be extracted from the flux-linkage waveform.)

At low speeds the motor does not have sufficient impedance or back-EMF to limit the current. The two main strategies are closed-loop current-regulation and voltage-PWM, as described in §§ 4.1-4.3 and Figs. 4.2-4.5 with both soft and hard chopping.

Depending on the states of the power transistors the source voltage applied to the phase winding can be V_s, $-V_s$, or 0. A truth-table defining the applied voltage as a function of the states of the switches is given below.

State	Q1	Q2	D1	D2	V
A	1	1	0	0	V_s
B	1	0	0	1	0
C	0	1	1	0	0
D	0	0	1	1	$-V_s$

In state A (both transistors conducting), the voltage equation for one phase is

$$V_s - 2v_q - 2r_q i = \omega\frac{d\psi}{d\theta} + Ri \qquad [17]$$

---

[3]A rigorous treatment would require that all loss mechanisms be implicit in the differential and algebraic equations describing the motor, the controller, and the supply. The separate, *a posteriori* calculation of "parasitic" effects such as losses is justifiable only if their effects are small in comparison with the main energy conversion. The nearest approach to a rigorous model is that of time-stepping finite-element analysis, but this is relatively cumbersome and slow in implementation, especially if the field equations are solved in three dimensions.

where v_q is a fixed voltage drop across each transistor, r_q is the resistance of each transistor, R is the phase resistance, i is the phase current, and ψ is the phase flux-linkage. The parameters v_q and r_q permit the modelling of most types of power devices (IGBT's, MOSFET's and bipolar transistors, mainly). If, as is normal, the resistances and v_q are small, this equation simplifies to

$$V_s \ = \ \omega \frac{d\psi}{d\theta} \qquad\qquad [18]$$

which is the differential form of equation 1. If V_s is assumed to be constant the flux-linkage ψ increases linearly with rotor position θ, as shown in Fig. 4.2. These conditions are maintained for $\theta_0 < \theta < \theta_C$.

At θ_C both transistors turn off and the voltage equation becomes

$$-V_s + 2v_d \ = \ \omega \frac{d\psi}{d\theta} + Ri \qquad\qquad [19]$$

where v_d is the forward voltage drop across each conducting diode. Again, if v_d and the phase resistance are small, this equation simplifies to

$$-V_s \ = \ \omega \frac{d\psi}{d\theta} \qquad\qquad [20]$$

which is the differential form of equation 3, showing that after commutation ($\theta > \theta_C$) the flux-linkage decreases linearly.

After integration of the voltage equation, the current waveform must be determined from the flux-linkage $vs.$ position waveform. Probably the fastest method for doing this is the elegant *analytical* method published by Ray and Davis [15]. In this method the magnetization curves are represented by the trapezoidal, piecewise-linear idealised inductance graph shown in Figs. 4.2-4.5, and saturation and fringing are ignored. Accordingly the magnetization curves have the general shape of Fig. 2.8. Although this is an oversimplification of the magnetization curves (neglecting both bulk saturation in the yoke and localised saturation at the overlapping pole-corners), the method delivers most of the important features of the current waveform, especially in relation to the development of controller architectures, and is easy to interpret physically.

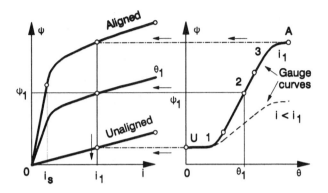

Fig. 4.6   Modelling of magnetization curves

For accurate work the magnetization curves must be modelled in such a way as to incorporate the important physical processes of *bulk saturation of the yoke* and *localised saturation near the overlapping pole-corners*. Indeed it has been shown in Chapter 2 that the most desirable form of the magnetization curves is closer to Fig. 2.7 than to Fig. 2.8. A further important effect is the *fringing field near the overlapping pole-corners*, and it is this that frustrates any approach based on piecewise-linear representation of the magnetization curves. The basic procedure for doing this was first laid out in [35].

PC-SRD uses two alternative *nonlinear* models for representing the magnetization curves: (1) cubic splines and (2) a piecewise combination of Fröhlich curves and straight lines with *gauge-curves*. The cubic-spline representation is very accurate. It models each magnetization curve by a series of up to 31 cubic splines, and there can be as many as 31 magnetization curves between the aligned and unaligned positions. The interpolation between these curves along the θ-coordinate is accomplished by a further orthogonal spline function that is re-constructed every time-step. The use of this representation permits dynamic simulation several orders of magnitude faster than would be possible with time-stepping finite-elements, but it is *still too slow* for rapid design and therefore there is provided the second model based on the piecewise combination of Fröhlich curves and straight lines with gauge-curves. This is shown in Fig. 4.6.

The unaligned curve is represented by a straight line. The aligned curve is represented by a straight line at currents below i_s, and by a parabola at currents above i_s. The unaligned and aligned curves are thus determined by only three points in the (i,ψ) plane, and they are represented by simple algebraic functions such that if ψ_A is known on the aligned curve, i_A can be calculated instantly. Similarly if ψ_u is known on the unaligned curve, i_u can be calculated instantly. Consider the problem of determining the current when the position and flux-linkage are known: say $\psi = \psi_1$ and $\theta = \theta_1$. The curve in the right-hand graph is a scalable "gauge" that fits between the unaligned and aligned curves in the left-hand graph. It is shown in the left-hand graph at the position of the current i_1. As the gauge slides right and left, its vertical scale changes along with the current and the vertical distance between the aligned and unaligned curves. On the right-hand graph, this change of scale can be seen by comparing the dotted curve with the solid curve. The solid curve fits between the aligned and unaligned curves in the left-hand graph at the current i_1. The dotted curve fits at a current less than i_1. The *shape* (mathematical function) of these gauge-curves is fixed; only the vertical scale changes. If θ is known, the ordinate $\theta = \theta_1$ can be drawn on the right-hand graph at a fixed location. Now the problem reduces to that of scaling the gauge-curve up or down until its intersection with $\theta = \theta_1$ occurs at the flux-linkage $\psi = \psi_1$. Since the scale of the gauge-curve is a simple function of the *parameter i*, it is relatively simple to program the gauging procedure to find i.

The speed of this procedure depends on the use of simple mathematical functions to represent both the gauge-curves and the aligned and unaligned magnetization curves in terms of the current i. In *PC-SRD*, the gauge curves are represented by a straight line **2** flanked by two Fröhlich curves **1** and **3** of the form

$$\psi = \frac{a\theta}{1 + b\theta},\qquad [21]$$

but other possibilities exist. A full account is given in [33] and [34], together with the methods used to calculate instantaneous and average torque.

The strength of this model is that it contains enough "structure" to model the effects of bulk and localised saturation and fringing around the partially overlapping pole-corners, but it is sufficiently simple to permit computation at the speed necessary for rapid, interactive design. In effect, it does for the switched reluctance motor what the assumption of a sinusoidal airgap flux distribution does for the induction motor. Its accuracy depends partly on the coefficients used in the gauge-curves, and partly on whether the gauge-curves have enough "structure" to cope with the variation of the magnetization curves between the aligned and unaligned positions.

Fig. 4.2 shows a typical single-pulse current waveform. In this example θ_0 coincides with the unaligned position where the phase inductance is low and varying only slowly. Consequently, at constant speed the current increases with an almost constant rate of rise $di/d\theta$. If the inductance is taken to be equal to the unaligned inductance L_u in this part of the waveform, and if resistance is neglected, the rate of rise with respect to rotor position is approximately

$$\frac{di}{d\theta} = \frac{V_s}{\omega L_u}.$$ [22]

Once the rotor and stator poles begin to overlap, the phase inductance begins to increase rapidly. Since $\psi = \psi(i, \theta)$

$$\frac{d\psi}{dt} = \frac{\partial\psi}{\partial i}\frac{di}{dt} + \omega\frac{d\psi}{d\theta} = l\frac{di}{dt} + e$$ [23]

where l is the incremental inductance (the slope of the magnetization curve at the position θ) and e is the back-EMF. In the switched reluctance motor e depends on the phase current and is zero if the phase current is zero. This contrasts markedly with the PM brushless motor, in which e is independent of the phase current. If the magnetization curves had the ideal shape of Fig. 2.7, and if they were evenly spaced in θ, then e would be constant throughout the torque zone from the unaligned to the aligned position. In practice this is not possible, but e can be substantially constant over a useful range of angles (the effective torque zone) if the phase current is constant.

Further insight into the nature of the back-EMF is obtained in the non-saturable motor, for which $e = i\omega dL/d\theta$. This shows that if the phase inductance varies trapezoidally with rotor position, as in the idealized trace in Figs. 4.2-4.5, then the back-EMF is proportional to the phase current. If the phase current is constant the back-EMF is also constant, and if i is equal to the rated current the speed at which $e = V_s$ is called the *base speed* [16].

4.5 Torque Waveform. If the two voltages on the right-hand side of equation 23 are multiplied by i, it appears that there are two power terms; in other words, the power obtained from the supply is partitioned. In the ideally saturable switched reluctance motor with magnetization curves as shown in Fig. 2.7, the term ei represents the electromechanical power conversion and the electromagnetic torque is just ei/ω throughout the angle for which e remains constant (the effective torque zone). This is virtually the same mechanism as in the DC or brushless DC motor. In effect, the small magnetizing current i_s sets up a fixed magnetizing MMF and the phase flux-linkage varies linearly with rotor position as the overlap between rotor and stator poles increases. If i_s is negligibly small, it is as if a magnet were rotating past the stator pole. This principle holds even if the magnetization curves in Fig. 2.7 have a finite slope, provided that they all have the same slope in the saturated region.

In a practical motor the term ei is not all converted to mechanical power: some of it is being stored in the magnetic field. The *worst case* is that of the non-saturable motor, for which the airgap power is

$$P_{gap} \;=\; T\omega \;=\; \frac{1}{2}i^2\frac{dL}{d\theta}\omega \qquad [24]$$

(neglecting losses), and the rate of storing energy in the field is

$$iL\frac{di}{dt} + \frac{1}{2}i^2\frac{dL}{d\theta}\omega. \qquad [25]$$

The sum of equations 24 and 25 is $i\omega d\psi/d\theta$. *Less than half* the supplied power is being converted into mechanical power.

This argument has been developed in terms of *instantaneous* torque and power, but it is equally important to consider the *average* torque, which depends on the areas on the energy-conversion diagram discussed in Chapter 2 [16].

The maximum energy ratio of the non-saturating motor is

$$E_{max} = \frac{\lambda - 1}{2\lambda - 1} \qquad [26]$$

and with $\lambda = 10$ this is 0.47. The maximum energy ratio for the ideally saturable motor of Fig. 2.7 is nearly 1, but in practice it is usual to find energy ratios of well-designed motors around 0.6-0.7.

The saturation which gives rise to the ideal magnetization curves in Fig. 2.7 is *local saturation* at the overlapping pole-corners, and this is a desirable feature because it maintains a fixed flux-*density* at the overlapping pole-corners. It is this that makes the flux-*linkage* increase linearly with rotor position as the rotor pole sweeps past the stator pole, as in a PM brushless motor. The bulk saturation of parts of the magnetic circuit remote from the airgap is always undesirable because it makes no contribution to keeping the flux-density constant in the overlap region and merely absorbs MMF.

Although the local saturation is beneficial, it is also desirable that the flux-density at which it occurs should be high, and with the smallest possible current i_s. This implies the need for both high permeability and high saturation flux-density B_s in the core steel. For normal electrical steels B_s is about 1.7T. Harris[4] has suggested that with this value of B_s, the *peak* torque generated at the onset of overlap is given by

$$T_{peak} = B_s r_1 L_{stk} \cdot 2 N_p i \qquad [27]$$

where N_p is turns/pole (with 2 stator poles active), r_1 is the stator bore radius, and L_{stk} is the stack length. Harris' formula is a simple means of calculating the peak torque available from a given switched reluctance motor. In a recent publication, Harris and Sykulsi [36] report that the linear variation of the (torque-producing) forces between the overlapping poles holds over a wide range of MMF.

The torque waveform can be calculated from equation [3] in Chapter 2, and this is done by a numerical procedure using either cubic-spline interpolation of a precalculated set of co-energy curves, with subsequent differentiation at constant current; or by the method described in [33].

---

[4] Harris, M.R. [1989] Private communication.

5 CAD of the switched reluctance motor

5.1 The Need for Computer Modelling. The switched reluctance motor operates in a series of strokes or *transients* and does not have a steady-state in which all its state variables are constant. In the DC motor the flux and current are normally constant in both time and space. In AC motors the same is true in a reference frame that rotates with the flux. But in no reference frame do the flux and current of the switched reluctance motor appear to be constant. This means that for all except the most basic sizing calculations, computer-based design methods must incorporate *simulation* capability as an integral part of the design process. In the design of conventional machines these have generally been regarded as separate tasks, often performed by different engineers. For the switched reluctance motor one integrated program must accomplish both tasks. An example, *PC-SRD*, is described in this chapter.

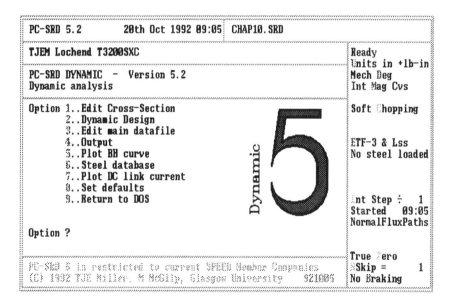

Fig. 5.1   Main menu of *PC-SRD*

5.2 PC-SRD.[1]   *PC-SRD* is a design and simulation program for switched reluctance motors and generators and their controls. The menu structure is shown in Fig. 5.2.

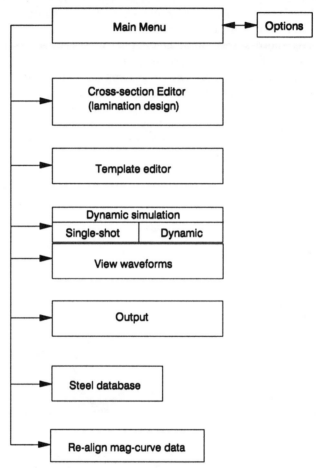

Fig. 5.2   Menu structure of *PC-SRD*

---

[1]PC-SRD is licensed via the SPEED Consortium at Glasgow University.

PC-SRD has two main functions: *sizing* and *simulation*. In *sizing*, a rough design is worked up from a simple specification, following principles described in Chapter 10, and then refined by progressive adjustment. The user specifies seven parameters to start the design:

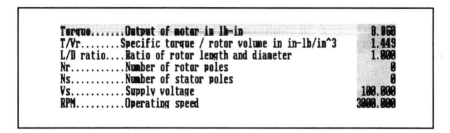

Entry panel for sizing parameters in PC-SRD

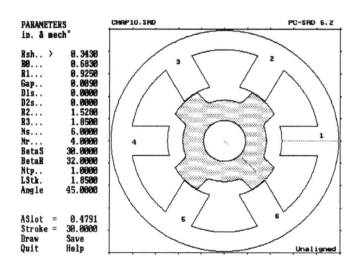

Fig. 5.3   Cross-section editor in PC-SRD

PC-SRD includes all of the tasks and resources involved in the preliminary design process. It cuts prototyping costs by producing good approximations at the first attempt, and it cuts engineering costs by providing a convenient learning and simulation facility.

Page 73

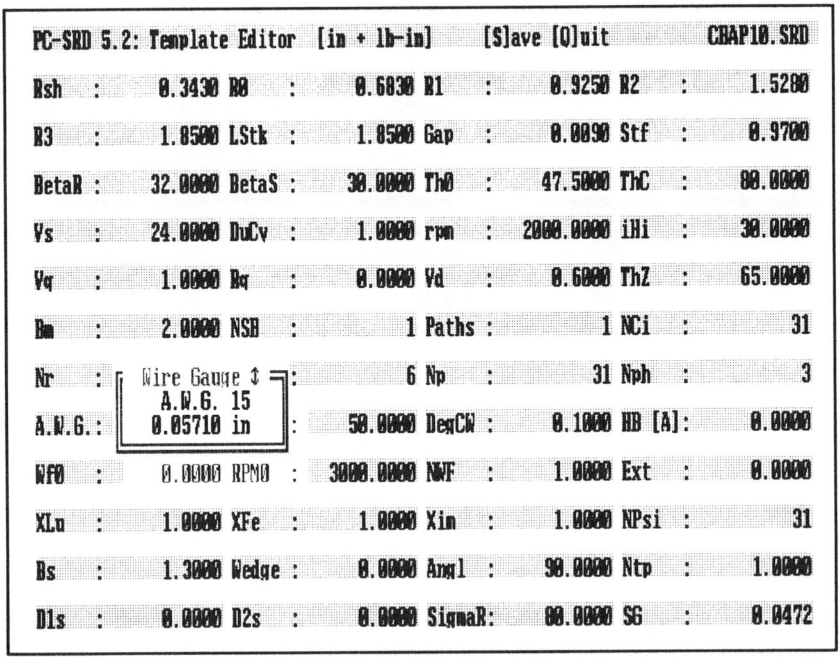

```
PC-SRD 5.2: Template Editor  [in + lb-in]      [S]ave [Q]uit          CHAP10.SRD

Rsh    :      0.3430 R0    :      0.6030 R1    :      0.9250 R2    :      1.5280

R3     :      1.0500 LStk  :      1.0500 Gap   :      0.0090 Stf   :      0.9700

BetaR  :     32.0000 BetaS :     30.0000 Th0   :     47.5000 ThC   :     80.0000

Vs     :     24.0000 DuCy  :      1.0000 rpm   :   2000.0000 iHi   :     30.0000

Vq     :      1.0000 Rq    :      0.0000 Vd    :      0.6000 ThZ   :     65.0000

Bm     :      2.0000 NSB   :          1 Paths :          1 NCi   :         31

Nr     :  ┌ Wire Gauge ↕ ┐:          6 Np    :         31 Nph   :          3
          │   A.W.G. 15   │
A.W.G. :  │ 0.05710 in    │:     50.0000 DegCW :      0.1000 HB [A]:      0.0000
          └──────────────┘
Wf0    :      0.0000 RPM0  :   3000.0000 NWF   :      1.0000 Ext   :      0.0000

XLu    :      1.0000 XFe   :      1.0000 Xin   :      1.0000 NPsi  :         31

Bs     :      1.3000 Wedge :      0.0000 Ang1  :     90.0000 Ntp   :      1.0000

D1s    :      0.0000 D2s   :      0.0000 SigmaR:     00.0000 SG    :      0.0472
```

Fig. 5.4   Template editor in *PC-SRD* (page 1 of 2)

Once the design is started, any parameter can be changed via the cross-section editor, Fig. 5.3, or the template editor, Fig. 5.4. Of the 70 input parameters, many can be left unchanged, but for detailed work there is wide scope for modifying designs and creating new ones. This form of CAD is faster, and more complete than "manual" design, which is extremely difficult for the switched reluctance motor. *PC-SRD* has an integral database of electrical steels, which can be extended or modified by the user. This database contains core-loss coefficients adapted for use with the non-sinusoidal flux waveforms.

In *simulation* mode, PC-SRD calculates and displays the current and torque waveforms, Figs. 5.5 and 5.6, and the energy-conversion loop, Fig. 5.7. All the calculated design and performance parameters are printed in the Output section, p.6. Details are contained in [31].

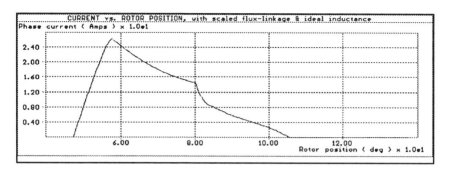

Fig. 5.5   Phase 1 current waveform (*cf.* Fig. 4.2). The data is for the example in Chapter 10.

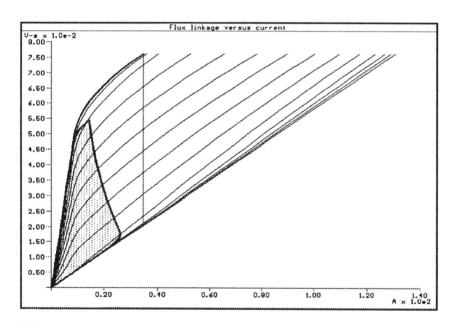

Fig. 5.6   Energy-conversion loop in *PC-SRD*

```
PC-SRD 5.2    TJEM Lochend T3200SXC                20th Oct 1992  09:34
Title   :    PC-SRD DYNAMIC  -  Version 5.2
SubTitle :   Dynamic analysis

Datafile :   CHAP10.SRD              Internal Mag Curves :    IPSI.PSI

DIMENSIONS [in] & [deg M]---------------------------------------------
Rotor  Rsh    0.343           Gap    0.009    Lstk              1.850
       R0     0.683  Phases  3                Stf               0.970
       R1     0.925  RPoles  4   BetaR 32.000 Ext               0.000
Stator R2     1.528  SPoles  6   BetaS 30.000 Aslot[in2]=       0.479
       R3     1.850  Ntp     1

WINDING DATA---------------------------------------------------------
Turns/pole       31    Temp.  [C]   50.000    Rph   [Ohm]=      0.111
Paths             1    Slot Fill =   0.331    Lau   [mH]=       5.731
NSH               1    M.L.T. [in]=  6.059    Lu    [mH]=       0.583
WDia  [in]= 5.710E-02  L/ends [in]= 75.331    Lau/Lu    =       9.835
A.W.G.           15    ACond [in2]= 2.561E-03 NwkPP     =          1

CONTROL DATA---------------------------------------------------------
Voltage      24.000    iHi   [A]    30.000    DuCy              1.000
rpm        2000.000    iLo   [A]    30.000    Dwell[degM]=     32.500
Th0 [degM]   47.500    Rq    [Ohm]   0.000    Dwell  [%]=       36.1
ThC [degM]   80.000    Vq    [V]     1.000    Strokes/rev=     12.000
ThZ [degM]   65.000    Vd    [V]     0.600    Ph.freq[Hz]=    133.333

PERFORMANCE---------------------------------------------------------
Torq[lb-in]=  11.035   WCu    [W]=  45.547    kVA/kW(rms)=      5.153
Power sh[W]= 261.170   WIron  [W]=   7.718    kVA/kW(pk) =     11.996
Effcy  [%]=   83.060   WWindage[W]= 0.000E+00 Deg. C/W          0.100
sigma [psi]=   1.108   Tot.Loss[W]= 53.265    TempRise[C]=      5.326

CURRENTS [A]----[PEAK]-----------------[MEAN]-----------------[R.M.S.]
IWpk     =    26.193   IWMean    =    7.916   IWrms     =      11.694
IQchop pk =   26.193   IQchop mean=   6.462   IQchop rms =     11.252
IQcomm pk =   26.193   IQcomm mean=   6.462   IQcomm rms =     11.252
IDchop pk =   13.778   IDchop mean=   1.454   IDchop rms =      3.188
IDcomm pk =   13.778   IDcomm mean=   1.454   IDcomm rms =      3.188
Jrms[A/in2]= 4.567E+03 DC     mean=   15.023*  DC  ripple =     5.256

IRON LOSSES [W]--[Eddy]----[Hyst]---[BMax]----[W/lb]
Rotor Yoke =    0.587    0.725   1.116    2.471         M19 26 gage
Rotor Poles=    0.239    0.286   1.488    4.318  1.5 W/lb 60Hz 1.5T
StatorYoke =    2.041    2.490   1.178    2.737         M19 26 gage
StatorPoles=    0.613    0.736   1.569    4.815  1.5 W/lb 60Hz 1.5T
Total      =    3.480    4.238   XFe     1.000

SUPPLEMENTARY OUTPUT------------------------------------------------
Wt Cu  [lb]=   0.927   Jm [ozfts2]= 1.118E-03 *DCsupply[A]=    14.935
Wt Iron[lb]=   3.271   Rho[Ohm-in]= 7.567E-07 ConvLoss[W]=     44.007
Tot.Wt [lb]=   4.198   Temp.fact. =  1.115    TR p-p [pu]=      1.556
Bm    [T]      2.000   Bs    [T]     1.300    Lau0  [mH]=       5.679
Psim [mVs]    76.138   Psis  [mVs]= 49.490    Lu0   [mH]=       0.492
im   [A]      34.681   is    [A]     8.635    Lss   [mH]=       0.194
Xim            1.000   XLu           1.000    ETF-3 & Lss
Th0B[degM]    45.000   ThCB[degM]   75.000    No Braking
--------------------------------------------------------END OF DESIGN
```

PC-SRD output data

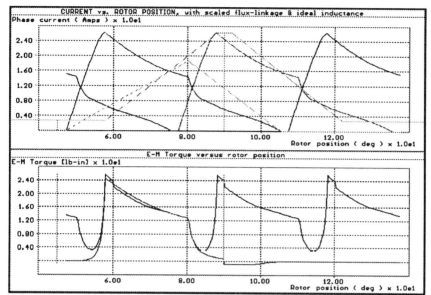

Fig. 5.7   Full set of waveforms, including three phase-currents, flux-linkage, idealised inductance, and torque.

5.3 Constant speed and varying speed. Most preliminary design work is centred on a fixed speed, or a small number of fixed speeds, even when the requirement is for operation over a wide speed range. In some applications, however, dynamic operation is of interest. For example, if the load is a reciprocating compressor the load torque varies widely during one revolution. The controller must cope with these variations. In other applications it is important to design controllers for rapid acceleration or speed reversal, without oscillations or overshoot.

To develop appropriate controllers to meet these needs, dynamic simulation software is needed, so that the controller architecture can be defined before detailed electronic design is finalized. Figs. 5.8 and 5.9 show dynamic simulations performed with *PC-SRD*. The motor model is the one described in Chapter 10, and the controller architecture is that of Fig. 7.7. The transient is defined as a step in reference speed (demanded speed) from 2000 to 3000rpm at *t* = 0.

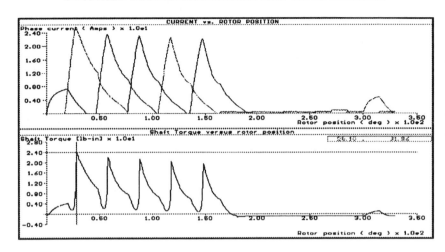

Fig. 5.8   Currents and torque during speed change. Note cursor
in lower graph.

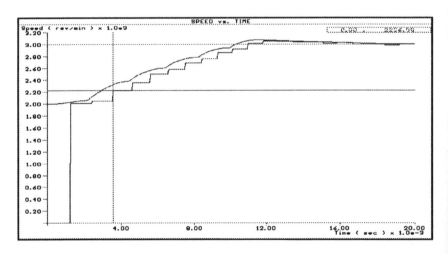

Fig. 5.9   Speed vs. time. Stepped graph shows encoder pulses.
Note cursor and speed reference line at 3000rpm.

The phase current waveforms in the upper graph in Fig. 5.8 are useful for determining the maximum peak currents in the motor windings and the controller transistors during transients. This is important because the currents during acceleration or speed reversal may need to be significantly larger than those at constant speed. Dynamic simulation software is as useful for sizing and rating the power semiconductors as it is for control system development.

5.4 Internal calculations performed by *PC-SRD*. The structure of *PC-SRD*'s internal calculations is shown in Fig. 5.10. It is based on the theory in Chapter 4 and [33]. The core operation is the integration of the differential voltage equation in each phase. This gives a new value of flux-linkage at the end of each integration step. The current corresponding to this value of flux-linkage is determined algebraically from the non-linear magnetization curves before the next integration step can be taken, as described in Chapter 4.

PC-SRD has a ranging facility that can be used to run a batch of up to 100 designs, with synchronized stepwise variation of any number of design parameters. The ranging facility permits the user to select any number of variables for plotting, and any one of over 100 performance parameters can be plotted against any design parameter. Because of the extremely nonlinear behaviour and the large number of variables, it is helpful to be able to design by directed search, and the ranging facility makes this possible under the supervision of the designer.

5.5 Finite-element analysis. Fig. 5.11 shows a detail of a finite-element mesh in which 9,276 elements are used to represent half the cross-section of a switched reluctance motor. The mesh is refined in the regions where the flux-density is expected to be high; where there is rapid spatial variation of the field; and in the airgap, which is represented by four concentric layers. In this example, the mesh also extends *outside* the frame, to include any fringing fields that leak outside when the frame is heavily saturated. The stator coil-side is represented by a simple geometrical shape, but for accurate work it is important to try to reproduce the exact cross-section of the coil, and even of each conductor within it, especially for the calculation of the total flux-*linkage* of the coil. The total flux-linkage is the sum of the flux-linkages of all the individual loops of wire, and these are

Page 79

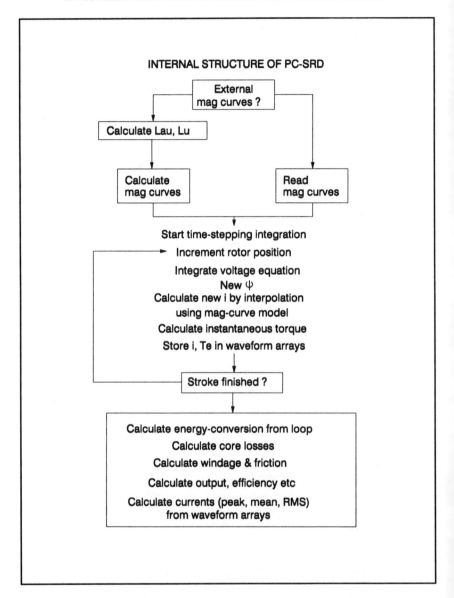

Fig. 5.10   Internal structure of *PC-SRD*

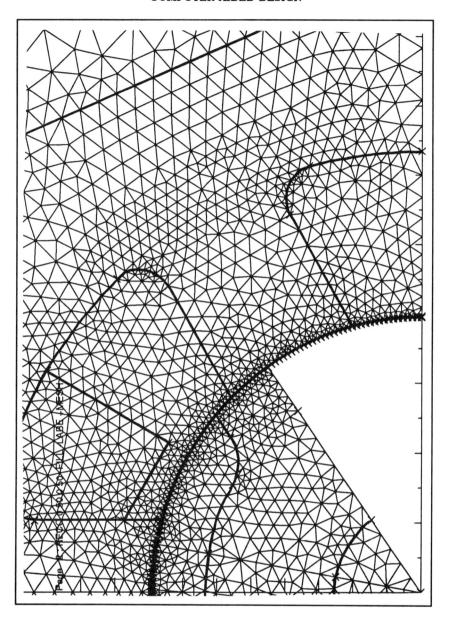

Fig. 5.11   Detail of finite-element mesh. (*Courtesy of W.L. Soong.*)

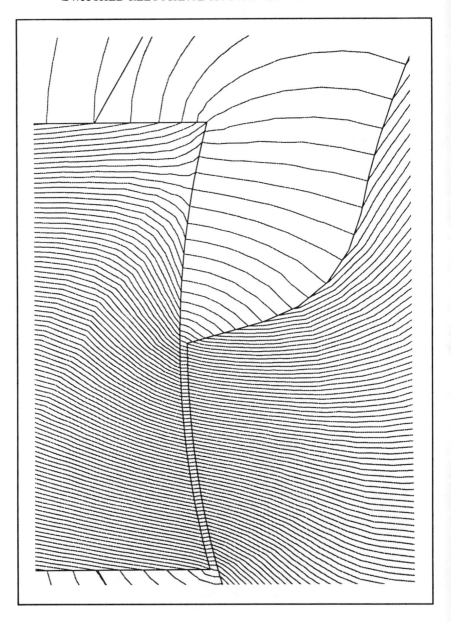

Fig. 5.12   Detail: finite-element flux-plot. (*Courtesy of W.L. Soong*).

generally not equal because the flux-density varies considerably across the cross-section of the slot, particularly in the unaligned position.

Fig. 5.12 shows a detail of the *flux-plot*, for a partial-overlap condition.

The accuracy of finite-element software depends on the skill of the user and on the nature of the problem. *Two-dimensional* solvers, for example, cannot be expected to give accurate magnetization curves for switched reluctance motors, especially in partially aligned position, because of the significant influence of end effects. However, 2-D solvers *can* be used very effectively to optimize *lamination geometry*. For example, it is a valid exercise to optimize the lamination geometry to achieve maximum inductance ratio for an infinitely long machine, and then apply that lamination geometry in a finite-length machine. For accurate results in partially overlapped positions, 3-D solvers are necessary.

A striking illustration of the potential of a 3-D solver has been given by C.W. Trowbridge in his book [37]. He reported the following results for calculated inductances:

Aligned		mH
Measured	2-D computation	3-D computation
66	66.5	68
Unaligned		mH
Measured	2-D computation	3-D computation
19.8	13.2	20.5

6. Power Electronic Controller

6.1 Unipolar operation. The direction of the torque does not depend on the signs or the values of flux-linkage and current, but only on the sign of $dL/d\theta$, the rate of change of inductance with rotor position. The advantage of this is that the flux-linkage and current may be unipolar (not alternating), and this mode of operation is preferred because it is believed to incur lower iron losses and to permit a simpler form of controller.

The controller must supply unipolar *current* pulses, precisely phased relative to the rotor position. It must regulate the magnitude and even the waveshape of the current, to fulfil the requirements of torque and speed control and to ensure safe operation of the motor and the power transistors. It must also be capable of applying pulses of reverse *voltage* for de-fluxing, as described in Chapter 4. Usually the voltage reversal is effected by freewheeling diodes.

Although these requirements are similar to those of AC or DC drives, many differences of detail preclude the use of an off-the shelf controller to control a switched reluctance motor.

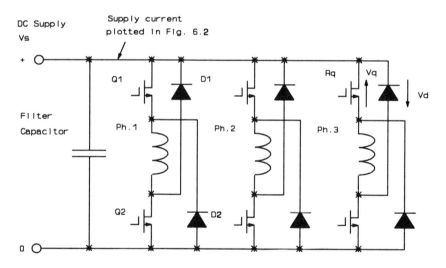

Fig. 6.1   Controller circuit with 2 transistors per phase. This circuit provides the maximum control flexibility and efficiency, with a minimum of passive components.

The most convenient power source is a battery or rectified AC supply. The ripple current tends to be large with an irregular waveshape rich in harmonics. The harmonics vary widely with speed and load, and it is essential to provide sufficient filter capacitance at the supply terminals, Fig. 6.1. Numerical simulation is needed to determine the ripple current under all conditions (including fault conditions).

6.2 Controller circuit. The highest efficiency, reliability, and control flexibility are achieved with derivatives of the circuit in Fig. 4.1, with an independent half-bridge for each phase. Although Fig. 6.1 shows m = 3 phases, any number can be used.

By controlling the upper and lower transistors in this circuit independently, all possible firing angles can be used, with soft chopping and with maximum regenerative braking capability and equal performance in forward and reverse directions. The current waveforms in various branches of this circuit are shown in Figs. 6.2 and 6.3 for soft chopping and voltage PWM.

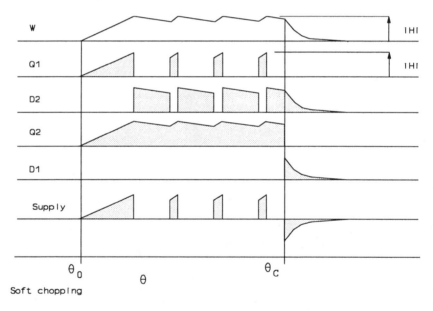

Fig. 6.2   Waveforms of winding (W), transistor, and diode currents during soft chopping in the circuit of Fig. 6.1.

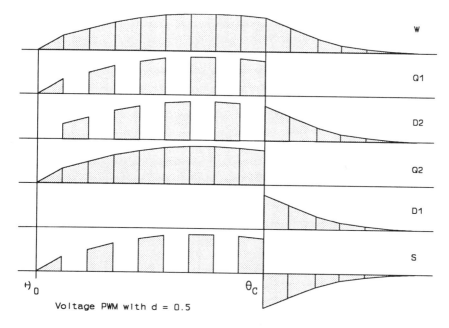

Voltage PWM with d = 0.5

Fig. 6.3   Waveforms of winding (W), transistor, and diode currents during voltage-PWM in the circuit of Fig. 6.1.

6.3 Number of switches per phase. Because the phase current is unipolar, there arises the possibility of controller circuits having fewer than two switches per phase. For 1-phase motors the minimum requirement is for one switch, but two-phase motors can, in principle, operate with only one switch if the second phase is in series with the freewheeling diode. For phase numbers of 3 or more, the minimum number of switches per phase for full control flexibility appears to be one. With a DC source, this must be a fully-controlled switch (transistor, GTO thyristor, or force-commutated SCR) because of the requirement to reverse the voltage at definite rotor positions and extinguish the flux each stroke. Attempts have been made to use an AC source with semi-controlled switches (naturally commutated SCR's), but the loss of control flexibility is severe.

For $m \geq 3$ the number of switches per phase therefore lies between 1 and 2.

Regardless of the number of switches per phase there must be a freewheeling path for the phase current. The circuit variants differ mainly in the details of this freewheeling path. Fig. 6.4 shows some well known examples. Only one phase is shown, but any number can be used.

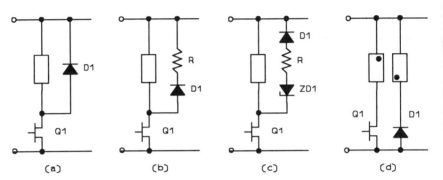

Fig. 6.4   Controller circuits with one switch per phase.

Circuit (a) with a plain freewheeling diode has no means of reversing the voltage at the winding terminals. Suppression of the flux therefore depends on the volt-drop in the diode and the winding resistance. Successful commutation is possible only with a high resistance, which is inefficient; or with early turn-off, which makes for a poor energy-conversion loop.

Circuit (b) uses external resistance to generate the suppression voltage (reverse voltage) across the winding, but this is inefficient and the suppression voltage decays with the current, decreasing the rate of fall of flux-linkage and lengthening the conduction angle with a "tail" that diminishes the efficiency and may delay the subsequent turn-on.

Circuit (c) uses a zener diode to sustain a higher value of suppression voltage during the freewheeling period.

Circuit (d) uses a bifilar-wound motor. When the transistor turns off, the current transfers to the secondary and freewheels through the diode. This circuit permits full voltage reversal but does not permit soft chopping because there is no "zero-volt" condition. The peak voltage on the transistor is twice the supply voltage, plus an additional transient voltage due to the imperfect coupling between the primary and secondary. This may be large enough to necessitate the use of a snubber, destroying the simplicity of the circuit and decreasing the efficiency, which is already seriously compromised by the use of a bifilar winding.

Fig. 6.5 shows the Oulton controller circuit.[1] This circuit has been used successfully in the well-known Oulton SR drive with 4-phase motors, with both GTO thyristors and IGBTs. It has the advantage of full reverse voltage for flux-suppression. The DC link voltage is split with capacitors. A disadvantage is that soft chopping is not possible, and balance must be carefully maintained between the phases, but the circuit successfully achieves the minimum of one transistor per phase without adding extraneous passive components or sacrificing control flexibility or efficiency.

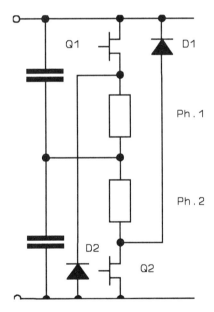

Fig. 6.5   Oulton type SR control circuit (only 2 phases shown)

---

[1] OULTON is a trademark of Graseby Controls Ltd, Lowestoft, England (formerly Tasc Drives). The OULTON SR drive is acclaimed for its high efficiency over a wide load range.

6.4 Controller circuits with shared components. Fig. 6.6 shows a three-phase controller circuit with four transistors. The upper transistor Q serves all three phases, while the lower transistors commutate the phases. Q must be operated with a sufficiently low duty-cycle so that when one phase is turning off, it has sufficient reverse voltage (averaged over several chopping cycles of Q). This circuit can be used with two or four phases also. A useful variant uses one upper between each pair of phases in a 4-phase drive, giving a total of 6 transistors, Fig. 6.7. If the chopping transistors are connected to phases 1 & 3 and 2 & 4 respectively, the problem of overlap is greatly reduced and all phases can have the maximum reverse voltage during turn-off. This circuit was published in [39], but it was independently developed by Prof. R. Krishnan of Virginia Polytechnic Institute [40]. An analysis of it appears in [41] along with several other related circuits.

The chopping transistor Q is working all the time, whereas the commutating transistors have a duty-cycle of only one-third. Consequently the switching and conduction losses as well as the mean and RMS currents in the chopping transistor are higher than in the commutating transistors. The use of a common chopping transistor also destroys the "fault tolerance" of the circuit in Fig. 6.1, because the phases are no longer as independent of one another.

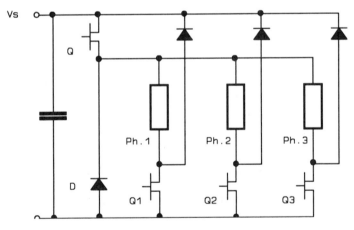

Fig. 6.6   (n+1)-switch circuit. The upper transistor serves all phases while the lower transistors commutate them.

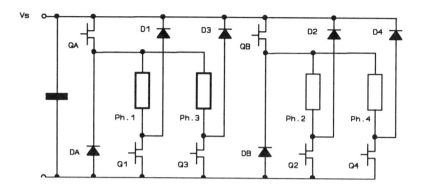

Fig. 6.7   2(n+1)-switch circuit: 6 transistors with 4 phases

Instead of sharing the supply transistors, the freewheeling circuit can be shared, as in Fig. 6.8, the C-dump circuit. All three phases freewheel into the dump capacitor C_d. The charge (and voltage) builds up on C_d, and is typically controlled to an average value of about $2V_s$ by a step-down chopper consisting of Q, L_d, and D. The circuit operates like a large snubber with energy recovery to the supply. The recirculating energy is generally a large fraction of the total throughput and therefore the chopper components and the dump capacitor are not small. Moreover, failure in the dump circuit is liable to permit the dump capacitor to charge up uncontrollably with catastrophic results.

A circuit of this type was reported by Byrne [44] in a drive system that was very nearly commercialized. Interestingly, Byrne had difficulty with torque ripple and adapted this controller for sinusoidal modulation of the phase currents, using the positive half-cycles only. He exploited the fact that in his motor the torque/angle curves were nearly sinusoidal, and by using the relationship

$$\sin^2\theta \;+\; \cos^2\theta \;=\; 1 \qquad\qquad [1]$$

he was able to reduce the torque ripple to a level of 5% (average to peak, open-loop).

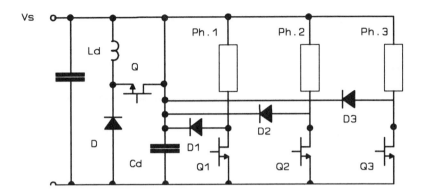

Fig. 6.8   C-dump circuit [43]

6.5 Generator circuits. The switched reluctance machine is capable of operating continuously as a generator. This is possible with those of the circuits reviewed above which can connect reverse voltage to the phase winding through freewheeling diodes. It must also be possible for the supply circuit to be capable of absorbing or diverting the returned power.

The switched reluctance machine can be caused to regenerate by retarding the firing angles so that the bulk of the winding conduction period comes after the aligned position, when $dL/d\theta < 0$.

In such cases excitation power is first supplied from the DC source while the transistors are conducting (i.e. during the dwell period from θ_0 to θ_C), and generating power is returned to the DC source during the de-fluxing or suppression period (from θ_C to θ_q). For generator action to be sustained, the energy returned during the de-fluxing period must exceed the excitation energy supplied during the dwell period. The difference between the electrical input energy and the electrical output energy is provided by the prime mover.

It is possible to separate the excitation circuit from the output circuit by means of a connection such as Fig. 6.9. This will be recognized as the circuit of the boost (or *up*) converter, and its operation is similar:

the only fundamental difference is that the inductance is not constant but varies with rotor position in such a way that mechanical work is added to the stored field energy during the conduction angle, so that the output power exceeds the excitation power.

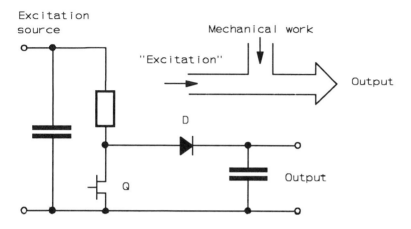

Fig. 6.9   SR generator circuit based on boost converter.

The output voltage in Fig. 6.9 must exceed the input voltage. This leads to a problem at low speed: the reverse voltage across the phase winding extinguishes the flux too quickly, and the energy-conversion loop is cut off with only a fraction of its maximum possible area. Fig. 6.12 illustrates this condition. The circuit of Fig. 6.10 alleviates the problem. It is based on the buck (or *down*) converter: the output voltage is less than the supply voltage. At a given speed the suppression angle $\theta_q - \theta_C$ is expanded approximately in the ratio V_o/V_s compared with the dwell angle $\theta_C - \theta_0$. Consequently a much fuller energy-conversion loop is obtained, as shown in Fig. 6.14.

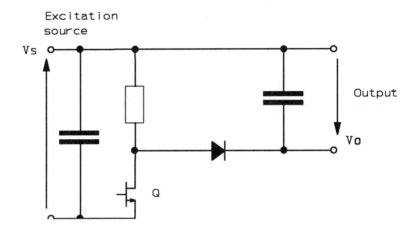

Fig. 6.10   SR generator circuit based on buck converter.

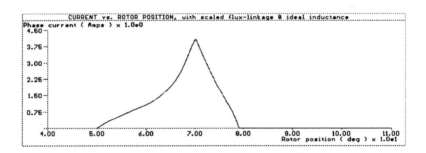

Fig. 6.11   Phase current in the generator circuit of Fig. 6.9

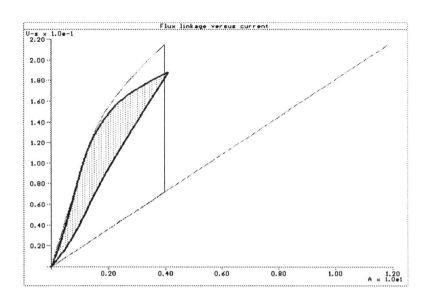

Fig. 6.12   Energy-conversion loop corresponding to Fig. 6.11. The flux is extinguished too quickly following commutation because the reverse voltage (i.e. output voltage) exceeds the supply voltage. Note that the loop is traversed in the clockwise direction.

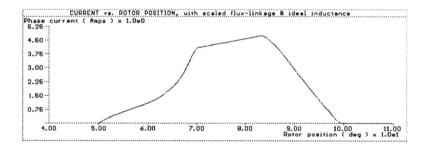

Fig. 6.13   Phase current in the generator circuit of Fig. 6.10.

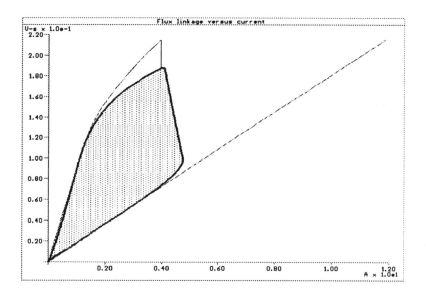

Fig. 6.14   Energy-conversion loop corresponding to Fig. 6.13. The low output voltage leads to a lower rate of flux-suppression, giving a fuller energy-conversion loop.

6.6 Excitation penalty. Because the switched-reluctance generator is a singly-excited machine, excitation energy must be supplied electrically every stroke. During the dwell angle, the excitation energy combines with the mechanical input energy to build up the stored field energy. No energy is supplied from this phase winding to the output during the dwell angle. After commutation, during the flux-suppression period, the stored field energy is released as output energy. At the same time, mechanical energy is converted directly to output energy, and augments the release of stored field energy to the output. The cycle ends when the flux reaches zero, and at this point the stored field energy is completely exhausted. If losses are neglected, the total output energy over each stroke exceeds the excitation energy by the mechanical work supplied.

Page 96

In a switched reluctance generator the relative size of these energy flows is especially important because the excitation power must be supplied locally from an independent source or by bleeding the output (which may require a separate DC/DC converter). By contrast, in a motor drive with an AC supply there is usually an unlimited supply of energy available to "stock" the DC link capacitor. Even so, the consequences of large amounts of recirculating or reactive energy are the same: increased power ratings for the switches and capacitors.

The energy flows in the switched reluctance generator can be characterized by an **excitation penalty** ε defined as

$$\varepsilon = \frac{P_{exc}}{P_{out}} \qquad [2]$$

where P_{exc} is the mean electrical excitation power and P_{out} is the mean electrical output power. In Fig. 6.10b ε is of the order of 60%. In Fig. 6.11b ε is of the order of 35%, which is better but still high. An ideal generator would have no excitation penalty, i.e. $\varepsilon = 0$, but this would require "square" magnetization curves involving ideally saturable material and zero airgap.

If the excitation power is supplied from an independent local source, the efficiency should be defined as

$$\eta = \frac{P_{out}}{P_{exc} + P_{mech}} \qquad [3]$$

with the excitation power counted as part of the input power. If the excitation power is bled from the output, the efficiency should be defined as

$$\eta = \frac{P_{out}}{P_{mech}}. \qquad [4]$$

7. Control System

7.1 Elements. Fig. 7.1 shows the block diagram of the switched reluctance motor connected to a load with closed-loop speed control. The controller structure is similar to that of AC and DC drives. The switched reluctance motor can be considered as a "black box" whose input is current and whose output is torque. Of course, "current" includes the currents of all phases, and it is understood that the current waveshape as well as its magnitude must be controlled to some extent because the waveshape is not pure DC or pure AC and indeed it varies with both speed and load. It may be necessary in some applications to control the waveshape of the current in a precise predetermined manner to minimize torque ripple.

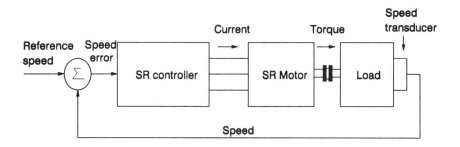

Fig. 7.1   Control system structure

The controller can be regarded as a black box whose input is the speed error and whose output is the motor current. The speed error is the difference between the reference (desired) speed and the actual (feedback) speed, which is derived from a speed transducer that is coupled to the load or the motor. Fig. 7.1 conveys only a basic example of closed-loop *speed* control. Similar diagrams can be put together for a *torque* controller or a *generator controller*.

7.2 LMB1008 Control IC. An example of a controller along the lines of Fig. 7.1 is the LMB1008 IC, an experimental device produced by National Semiconductor in collaboration with the SPEED Laboratory. Fig. 7.2 shows the structure of the LMB1008, which is implemented in a 28-pin, 24V integrated circuit.

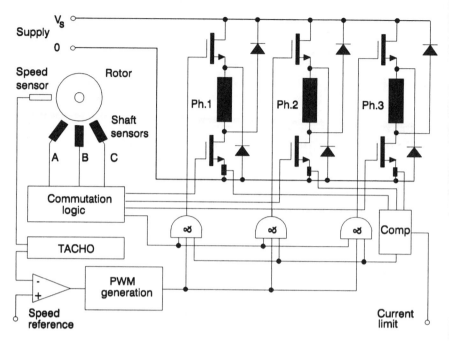

Fig. 7.2   LMB1008 Controller

The speed error amplifier produces a pulse-train whose duty-cycle is proportional to the speed error, and this is used to chop the external power transistors at a fixed frequency. Regulation is therefore by voltage-PWM rather than current-regulation (see Chapter 4); but a current-limit input is compared with external current sensor signals to turn the transistors off if the current exceeds a preset value.

Fig. 7.3 shows a block diagram of the LMB1008 IC. The LMB1008 can control any 3-phase SRM. Commutation is based on a shaft-position signal consisting of three sensor waveforms shown in Fig. 7.5. Fig. 7.4 shows a slotted disk for use with optical interrupters, but Hall-effect and other types of sensor are used. The function is the same as in the squarewave type of brushless DC permanent-magnet motor drive, but it is probably fair to say that the switching precision is more critical in the switched reluctance drive, especially at high speed where the current waveform is critically dependent on the firing angles. Precision of 0.5°, or even 0.25° is desirable.

Page 100

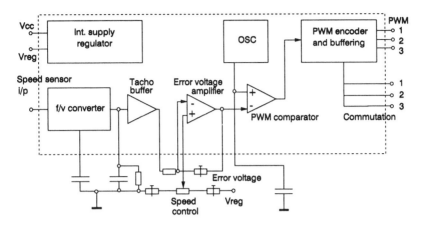

Fig. 7.3   Block diagram of LMB1008 controller

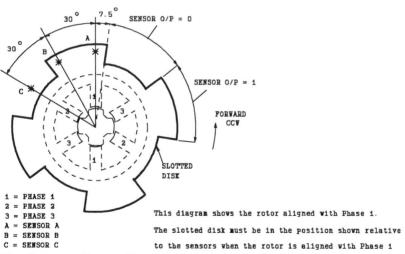

1 = PHASE 1
2 = PHASE 2
3 = PHASE 3
A = SENSOR A
B = SENSOR B
C = SENSOR C

This diagram shows the rotor aligned with Phase 1.

The slotted disk must be in the position shown relative

to the sensors when the rotor is aligned with Phase 1

Fig. 7.4   Generation of shaft position signal

Fig. 7.5 shows some of the commutation waveforms used in the LMB1008. Firing angles can be selected by combinatorial logic from the A,B,C, position sensor signals. Several "modes" are available. In terms of a 6/4 motor, where the aligned position is at 90° and the unaligned position is at 45°, the main modes are

Normal	52.5°/82.5°
Boost	37.5°/67.5°
Advanced	22.5°/67.5°
Braking	82.5°/112.5°

Normal mode is used at low speed; boost mode at high speed; and advanced mode at very high speed. This discrete selection of firing angles has the advantage of simplicity and it provides a wide speed range, with symmetrical operation in both directions. Fig. 7.5 shows typical phase current waveforms.[1] The voltage PWM can be seen in Fig. 7.5(b) and (c); and current regulation in (a).

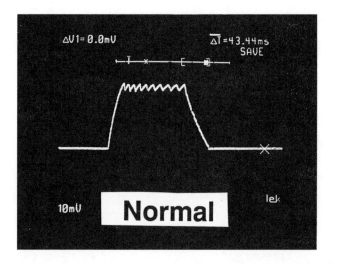

Fig. 7.5(a) Phase current waveform; *NORMAL* mode; current regulation

---

[1]Scope photos in Fig. 7.5 were prepared by Calum Cossar.

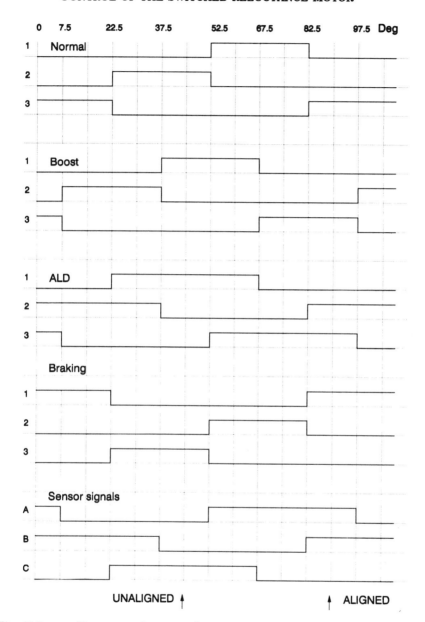

Fig. 7.6   Commutation waveforms in LMB1008, showing some of
the firing-angle combinations

Page 103

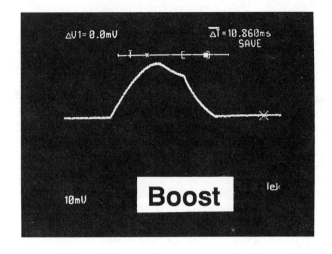

Fig. 7.5(b)   Phase current waveform; *BOOST* mode; voltage-PWM

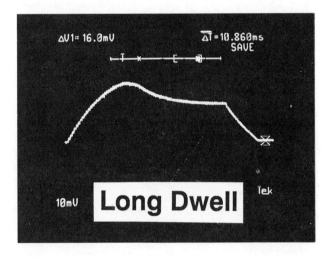

PWM

7.3   General controller architecture.   While it is easy and inexpensive to make a basic switched reluctance controller in the manner of the LMB1008, it is much more challenging to meet the more general requirements of four-quadrant operation (i.e., forward/reverse and positive or negative torque), with "seamless" transition between quadrants. True "servo quality" control imposes the further requirements of very low torque ripple, rapid dynamic response, good stability, the ability to operate at zero speed, and smooth reversing. Even without these servo-quality requirements, optimized performance for simple variable-speed drives requires continuous control of the firing angles.

The DC commutator motor and the brushless DC motor are well adapted for these advanced requirements because their torque is proportional to current. With vector control (field-oriented control), AC induction motors and PM synchronous motors effectively acquire this characteristic of the DC motor. In theoretical terms, this is possible because of the fact that the equations of the AC motor can be transformed into those of the DC motor by means of reference-frame transformations (the dq-axis transformation).

The switched reluctance motor has no "dq-axis transformation", and no "field oriented control" principle has been developed for it. Therefore, the control requirements of four-quadrant operation and servo performance can be met only by high-speed real-time controllers which operate with phase currents and voltages directly, and not with slowly-varying dq-axis quantities. Such high-speed control is used in advanced DC and AC drives already, in order to achieve the highest dynamic performance. What makes the switched reluctance motor different is that the relationships between torque, current, speed, and firing angles are highly nonlinear and vary as functions of speed and load.

A more general switched reluctance controller architecture is shown in Fig. 7.7. Although this is more complicated than the LMB1008, it still falls short of achieving "servo-quality" performance because of the absence of any control means for profiling the current waveforms to eliminate torque ripple, and the absence of any means for compensating the nonlinearities in order to achieve constant torque/amp. The controller of Fig. 7.7 can be implemented in any one of a wide variety of ways, including single-chip microcontrollers.

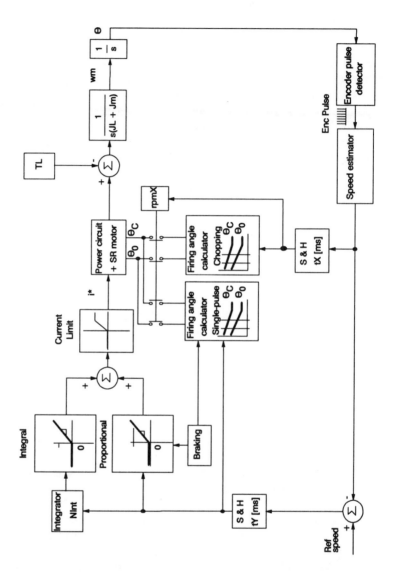

Fig. 7.7   Controller   architecture   providing   four-quadrant
           operation, closed-loop speed control, and firing-angle
           control. This controller is not sufficient for servo-quality
           performance. [31]

In Fig. 7.7 the switched reluctance motor and its power electronic controller circuit are shown as a "black box" whose output is torque, and whose inputs are the current demand signal i^* and the turn-on and turn-off angles θ_0 and θ_C. The black box is assumed to include a current-regulator that can impose a current waveform like the one in Fig. 7.5(a). The diagram also includes the notional integration of the torque and velocity equations showing the true speed and rotor position. The rotor position is sensed with an encoder (or resolver) which generates a pulse train. From the pulse train the speed is estimated by a digital algorithm.

The digital speed estimate is compared with the demanded speed (reference speed), and the error is applied to the current-regulator which generates the current demand signal i^*. Both proportional and integral gains are shown.

If the speed exceeds the reference speed, true "four-quadrant" operation of the type required in servo drives requires that the motor produce a braking torque. In many plain *variable-speed* applications this is not necessary because the load torque causes sufficient deceleration whenever the motor torque is decreased below it. To produce a braking torque, the current must remain finite while the firing angles are retarded. The amount of braking torque is a nonlinear function of the current and the firing angles, as is the motoring torque.

In Fig. 7.7 braking torque is produced by a sudden shift in the firing angles whenever a negative speed error is detected. This is illustrated in Fig. 7.8. At the "initial speed" shown, the firing angles are θ_0 and θ_C. As the speed increases, these angles are both advanced until the speed reaches the reference speed. At this speed the firing angles are at the maximum advance, θ_{0R} and θ_{CR}. The turn-on angle is advanced faster than the turn-off angle. If the speed exceeds the reference speed (demanded speed), the firing angles are suddenly retarded to the fixed values θ_{0B} and θ_{CB} and braking torque is generated. Braking torque remains under the control of the current-regulator, which continues to control the current in accordance with the speed error, provided that the speed is sufficiently low for the current-regulator to work. For braking from very high speeds this will not be possible and the braking angles will need to be modulated according to the speed error.

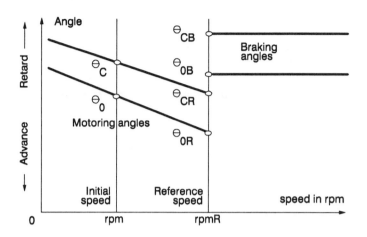

Fig. 7.8   Firing angle variation with speed.

In motoring, control by current regulation at higher speeds is either impossible or unsatisfactory, and firing-angle control must be used to control the torque. The generation of a mapping of firing angles as functions of speed and/or speed error is a nonlinear and complicated process. Because of the nonlinearity, it would appear that the only method available is to conduct an exhaustive series of dynamometer tests, in which the optimum firing angles are painstakingly sought by trial and error. The same process could be followed using a numerical simulation, for example with PC-SRD. The required mapping of firing angles is similar to that which is used in engine-control computers in motor-cars.

7.4   Sensorless Control.   Although stepper motors operate successfully without shaft position feedback, this "open-loop" mode of operation is not stable in the switched reluctance motor. Particularly in the United States, great efforts have been made to devise controllers for switched reluctance motors that operate without shaft position sensors - "sensorless control", and several ingenious schemes have been patented and published. The main reasons for eliminating the shaft position sensor are as follows

1.   to reduce cost;

2.   to increase reliability and adapt the drive for operation
        in harsh environments, such as high-temperature
        environments, or in hermetic compressors.

It may seem ironic that the motor itself is so simple and rugged, yet
it depends on accurate shaft position information for anything
approaching "optimum" performance. However, Lang [52] has pointed
out that a motor may make a very good sensor: the idea of
integrating the sensor with the motor is not just a question of
eliminating the sensor, but of investigating the possibility of an
overall improvement in the system performance. In his work on
observers he set out to achieve this, and reported a resolution of
1/50,000th of a revolution obtained with an observer whose input
signals were just the phase voltages and currents. Resolution of this
quality is sufficient for direct-drive robot motors.

Sensorless control schemes can be divided into four classes:

1.   Open-loop control with some form of additional
        stabilization

2.   Passive waveform detection

3.   Active probing

4.   State observers.

1. Open-loop control with stabilization.[43] When a shaft position
sensor is used, the independently controlled variables are the angles
θ_0 and θ_C, together with the current or the voltage. Essentially this
makes the innermost loop a torque-control loop, and the acceleration
is determined by the difference between the motor torque and the
load torque. A closed speed loop is usually added as an outer loop,
and the synchronization of the current pulses is maintained at all
times.

In open-loop control, the dwell angle $\theta_D = \theta_C - \theta_0$ (or more precisely,
the dwell *period*) is controlled, but there is no position sensor to
synchronize either θ_0 or θ_C to the rotor position. Instead, the

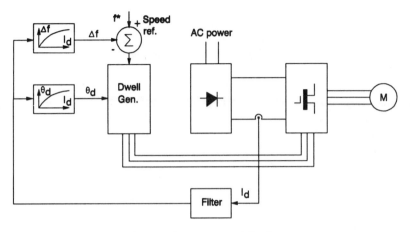

Fig. 7.9   Stabilized open-loop control [55]

commutation frequency is controlled. Under certain conditions the motor will run synchronously at the correct speed, exactly like a stepper motor. Although the idea is simple, switched reluctance motors are unstable in this mode because of the very large stroke (step) angle and poor damping (which is partly a result of their being designed for high efficiency). The scheme can be improved by adding stabilizing circuitry, the basic idea being to detect any departure from stable synchronous operation and take corrective action. Of course, it is assumed that the motor has been able to start and run up to synchronous speed in the first place. Bass stabilized his controller by increasing the dwell angle in response to any sudden increase in DC link current, which is taken to indicate an increase in load torque [55]. Stable operation was achieved with a small motor intended for use in driving a refrigerator condenser fan. The advantage of Bass' scheme is low cost, but it has to be "tuned" for the individual application and would not be suitable for compressor drives because of the extreme variation of load torque. There is no explicit signal or measurement of shaft position in this scheme.

2. Passive waveform detection. This scheme relies on natural points of inflection, or maxima or minima, in the phase current waveform as the rotor passes certain positions such as the aligned or unaligned position. Early publications of such schemes include [56,57] and were originally developed for closed-loop operation of stepper motors.

Page 110

3. *Active probing.* The principle is shown in Fig. 7.10. A squarewave voltage is applied to an *unexcited* phase. The resulting current waveform is as shown. The current pulses increase as the phase inductance decreases, reaching a maximum at the unaligned position. Thereafter they decrease as the inductance increases. By detecting when the current pulses fall below a threshold, a particular rotor position can be detected and used immediately for switch-on or commutation of the power transistors. The peak value of the current pulse is given by the equation [53]

$$i_{peak} = \frac{VT}{L(\theta)}$$

where T is the duration of the voltage pulses, V is the applied voltage, and $L(\theta)$ is the inductance at the rotor position θ. Several schemes of this type have been reported [53,58-60], including variants with compensation for variations in parameters such as speed and supply voltage. Harris and Lang [53] give an interesting analysis of their scheme, including the effects of eddy-currents in the motor laminations, mutual inductance, inverter noise, and digital quantization error. They describe a controller with 1% position error that used current pulses less than 10% of rated current, based on an 8031 microcomputer with 8-bit A/D converters, and they showed examples of operation at 1500rpm with a probing frequency of 8kHz.

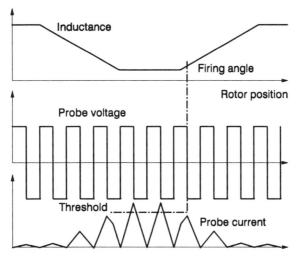

Fig. 7.10   Active probing of unexcited phase [53,54,58-60]

4. Observers. Lumsdaine and Lang [52] described a state observer which is effectively a mathematical simulation of the motor running on-line on a microcomputer in parallel with the actual drive. Measured currents and voltages from the actual drive are fed to the model, which is formulated in such a way that explicit signals representing rotor position and speed are available from it. They presented an analysis of the observer's stability and its convergence, together with experimental evidence obtained with a 242-pole servo motor, for which they were able to extract position information accurate to 1/50,000th of one revolution. The settling period of the model was approximately 2 electrical revolutions or 2.5 degrees of rotation at an angular velocity of 1.62 rad/s (15rpm). Although the observer could be said to be "computationally intensive", requiring a 32-bit processor and other high-speed integrated circuits, for low-speed applications the approach seems very promising especially in view of steady cost reductions in digital signal processors (DSP). Lumsdaine and Lang showed that operation is possible at zero speed provided that at least two phases are excited.

7.5 Future work in switched reluctance motor control. The most advanced forms of switched reluctance control have always had the ability to vary the firing angles as the speed and torque vary. Since the performance is sensitive to the precise values of these angles, accurate shaft position feedback is necessary if performance is to be maximized, and the shaft position signal must be immediately available without delay to be used for commutation.

To obtain smooth torque, the current waveforms must be forced into a predetermined waveshape which has to be stored in electronic memory and then injected into the phase windings by means of a high-bandwidth current regulator. The principle is no different from that of the field-oriented controller used with AC motors, except that the AC motor currents are sinusoids. The sinewaves simplify the design; but more importantly, they make the field-oriented controller relatively independent of the particular induction motor it is controlling. To perform at this level, every switched reluctance motor must have its own "personal" current-profiling circuitry. Of course it is technically feasible, but the number of applications for which this approach is economical is bound to be limited, in spite of the apparently low cost of the switched reluctance motor itself.

A significant breakthrough would occur if control theory could be developed to determine the optimum current waveforms of the switched reluctance motor automatically in real time, avoiding the need for pre-programmed "personality IC's" to be painstakingly prepared for every application. Even if a torque observer could be developed, it would effectively solve the problem by enabling closed-loop torque control within the speed loop. These possibilities have been recognized for some time [51,61], and significant efforts have been made to demonstrate the capability of the switched reluctance motor to produce very low torque ripple [62-64].

7.6 Fault tolerance. The switched reluctance motor and its controller are perceived to have a degree of fault tolerance not found in other motor drives, stemming from the lack of excitation and the independence of the phases. A detailed account of various faults and means for protection has been given by Stephens [65]. Because there is no field winding or permanent magnet, an open-circuit condition in the phase circuit produces no generated voltage, and a short-circuit condition produces no fault current. Furthermore, the outage of one phase does not destroy the rotating MMF as it does in an AC motor, and the other phases can continue unaffected. Another advantage is that controller circuits of the form of Fig. 6.1 do not have a shoot-through path. In AC inverters, if both the upper and lower transistors in one phaseleg are switched on simultaneously, the DC supply is shorted, but in the switched reluctance motor controller, this possibility is effectively eliminated unless the phase winding itself is short-circuited.

8 Losses and Cooling

8.1 Copper Losses. These can be calculated as for any other motor from the I^2R products for all the phase windings, where R is the effective resistance per phase. R is greater than the DC resistance (typically by a few percent) because of both skin- and proximity-effects, and of course it increases with temperature at the rate of about 20% for every 50°C rise in the case of copper windings. Skin effect is likely to be a problem in very high speed motors wound with small numbers of turns, and it may be necessary in such cases to use stranded conductors or *Litz wire*.

8.2 Excitation Loss Penalty.[66] The excitation power is supplied through a component of phase current. In permanent-magnet motors the excitation is supplied by the magnet. The reluctance motor therefore has an inherent disadvantage with regard to its efficiency since the excitation component in the phase current increases the copper losses and the losses in the power transistors. This *excitation loss penalty* is the price paid for not using magnets.

The excitation loss penalty is a function of physical size. In small motors the effect of iron saturation is negligible and the mechanical output power P_m is determined by the product of some representative current i and an inductance-difference parameter ΔL:

$$P_m = i^2 \Delta L \qquad [1]$$

where ΔL has dimensions $\mu_0 N^2 s^2 / g \sim s^2 / g$ and s is a generalized linear dimension representing the *scale* or size of the motor and g represents the mechanical airgap between stator and rotor. Note that s and g do not necessarily scale together in small motors, because g may reach some minimum manufacturable level as s is decreased, to allow for clearance and tolerances. The copper loss is represented by an equation of the form

$$P_{Cu} = i^2 R \qquad [2]$$

where R has dimensions $\rho s / s^2 \sim s^{-1}$. Assume that as the scale s is changed, the current level is adjusted so that the copper losses remain proportional to the area of surfaces available for cooling. Then $P_{Cu} \sim s^2$ and

$$i^2 R \sim \frac{i^2}{s} \sim s^2 \tag{3}$$

so that

$$i \sim s^{3/2}. \tag{4}$$

Under this constraint the scaling of the mechanical power is given by

$$P_m \sim i^2 \Delta L \sim s^3(s^2/g) \sim s^5/g. \tag{5}$$

The per-unit copper loss is therefore given by

$$\frac{P_{Cu}}{P_m} \sim \frac{s^2}{s^5/g} \sim \frac{g}{s^3}. \tag{6}$$

If g scales with s, then P_{Cu}/P_m scales with s^{-2}, but if the gap is fixed at some minimum-clearance value then P_{Cu}/P_m scales with s^{-3}. In the PM motor, the mechanical output power is given by

$$P_m \sim i\phi \tag{7}$$

where ϕ is a flux paramter and $\phi \sim Bs^2$. The flux-density B is not sensitive to the airgap when hard magnets are used, and therefore B is independent of scale in small motors. Hence

$$P_m \sim is^2 \sim s^{3/2}s^2 \sim s^{5/2}. \tag{8}$$

The per-unit copper loss is therefore given by

$$\frac{P_{Cu}}{P_m} \sim \frac{s^2}{s^{5/2}} \sim s^{-3/2}. \tag{9}$$

Comparing the scaling of P_{Cu}/P_m in the reluctance and PM motors, it is clear that in the reluctance motor the per-unit copper loss increases at a faster rate than in the PM motor as the scale is reduced.

8.2 Core losses. From what little detailed test data has been published, it appears that the core losses in switched reluctance motors are relatively low, even though the switching frequency is higher than in AC motors of the same speed and comparable pole number, and the flux waveforms in various parts of the magnetic

Page 116

circuit are decidedly nonlinear. The reasons for this were discussed briefly in [67]. The magnetic loading in the switched reluctance motor tends to be low, while the electric loading tends to be high, relative to the equivalent values in AC motors. Moreover, the volume of iron is lower. In very high-speed applications, the core losses grow rapidly and may become the dominant component of losses.

The core losses are generally calculated using the Steinmetz equation,

$$P_{Fe} = C_h f B_{pk}{}^{a + bB_{pk}} + C_e f^2 B_{pk}{}^2 \qquad [10]$$

where C_h and C_e are the coefficients of hysteresis and eddy-current loss, and a and b are constants. Manufacturers' loss data is usually presented in the form of graphs showing the *total* core loss as a function of frequency and flux-density, and standard values are quoted for 1.5T at 50 or 60Hz with sinusoidal flux-density.

In the switched reluctance motor the flux waveforms in different parts of the magnetic circuit are non-sinusoidal. Fig. 8.2 shows the construction of flux waveforms for a 3-phase 6/4 motor, given by Lawrenson in [2]. The construction of these waveforms has also been described by Krishnan in [68].

In order to deal with the non-sinusoidal flux waveforms, *PC-SRD* uses the following procedure [31]. First, the eddy-current term in the Steinmetz equation is re-written in terms of dB/dt:

$$P_{Fe} = C_h f B_{pk}{}^{a + bB_{pk}} + C_{e1}\left[\frac{dB}{dt}\right]^2 \qquad [11]$$

where $C_{e1} = C_e/2\pi^2$. The coefficients C_e, C_h, a and b are extracted from loss curves using a curve-fitting procedure. Then, the flux waveforms in the different parts of the magnetic circuit are calculated, and the mean value of $(dB/dt)^2$ is calculated for each section. The losses are then calculated from equation (11) on a per-kg basis, and multiplied by the weights of iron of the respective sections. Fig. 8.1 shows that some sections do not experience alternating flux, but unipolar flux. In these cases the hysteresis loss term is reduced by a factor of 2 or 3 to allow for the "minor loop" effect in the hysteresis loop. A similar approach has been published more recently by Slemon in connection with brushless PM motors [69].

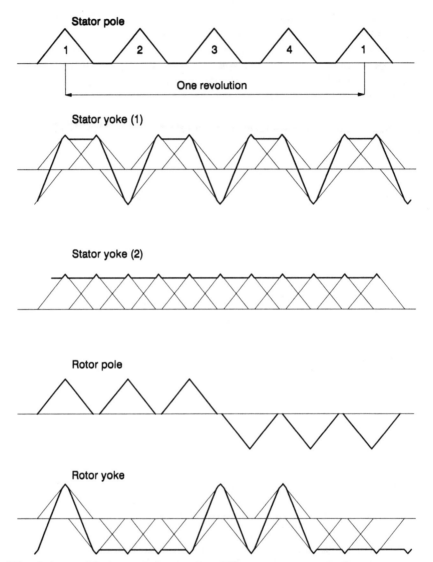

Fig. 8.1   Flux waveforms in different parts of the magnetic circuit published by Lawrenson [2] for a 3-phase 6/4 motor. Note that different sections of the stator yoke have different waveforms according to the winding polarities.

8.3 Heat. Heat transfer is at least as important as any aspect of the electromagnetic design, yet *almost all* textbooks on electric machinery leave this topic out of consideration. The analysis of heat transfer and fluid flow in motors is actually more complex, more nonlinear, and more difficult to analyze than the electromagnetic behaviour. It is often dealt with by means of rules of thumb and crude equivalent circuits, and rarely receives the "high-tech" attention lavished on the electromagnetic aspects.

Perhaps there is some justification for using approximate methods for heat transfer, when exact methods are pursued for electromagnetic design. The electromagnetic design determines the geometry of laminations which are cut to very fine tolerances. Their geometry and thickness, together with their material properties and the design of the windings, determine whether or not the motor will deliver the required torque. They also determine the precise voltages and currents that will be experienced by the power semiconductors in the controller. All of these items critically affect the manufacturing cost. By contrast, as long as the temperature rise does not exceed a nominal or specified value, the actual thermal condition of the motor mainly influences how long the motor will last, and has only a marginal influence on whether the torque can be delivered. Furthermore, the motor designer may have little control over the ultimate thermal environment of the motor, so there may be little point in attempting exact thermal analysis.

There are two major aspects to the thermal problem.

First is the question of heat removal. In most motors this is achieved by a mixture of air convection, conduction to the frame mountings, and radiation. In highly-rated machines direct cooling by oil mist or even liquid coolants can be used to achieve high power density. The heat-removal can be approximately analyzed with simple formulas from the theory of heat transfer and fluid dynamics.

Secondly, there is the question of the temperature distribution *within* the motor. This is essentially a diffusion problem. It is difficult to deal with theoretically, because of three-dimensional effects and "imponderable" parameters such as the thermal contact resistance between, say, a bunch of copper conductors and a slot liner. Empirical rules are available, to be used with care. The temperature distribution problem is essentially the problem of finding the hottest temperature in the motor, given a certain distribution of losses and a known rate of heat removal. Of course, the steady-state temperature distribution is *enormously* different from the transient distribution, and different methods of analysis are needed for the two cases.

The main reasons for limiting the temperature rise of the windings and frame of a motor are:

1   to preserve the life of the insulation and bearings;

2   to prevent excessive heating of the surroundings; and

3   to prevent injury caused by touching hot surfaces.

The "life" of electrical insulation can be predicted only by statistical methods, but in broad terms the life is inversely related to the temperature, and the relationship is exponential, so that a sustained 10°C increase in temperature reduces the insulation life by approximately 50%. Intermittent periods at higher-than-normal temperature can be tolerated repeatedly, depending on their duration and the actual temperatures reached. A spectacular example of this is the *FUMEX* motor[1], which is used to extract fumes via the ventilation systems of public buildings and concourses in the event of fire; these motors can operate in an ambient temperature of 300°C for a limited period of 30 minutes. (After that the insulation life is well-and-truly used up). Similar considerations apply to bearings. Grease-lubricated bearings may be filled with high-temperature grease for hot-running applications, but in aerospace machines the bearings are usually lubricated by oil which is separately cooled.

Heating of the surroundings is obviously undesirable especially if the motor is heating the equipment it is driving. For this reason it is important to minimize rotor losses, which are difficult to remove and are conducted along the shaft. Switched reluctance motors have cooler rotors than DC or induction motors, but not as cool as those of brushless DC PM motors.

---

[1]*FUMEX* is a trademark of Brook Crompton Parkinson Motors Ltd., England.

In some applications such as hermetic compressors used in air-conditioning, refrigeration, etc., the motor losses are removed by the working fluid, reducing the thermodynamic efficiency of the system.

Exposed surfaces must be kept below 50°C (actual temperature) to avoid injury or harm to people. In certain applications, for example under-bonnet automobile auxiliaries, this requirement is impossible to meet because the *ambient* temperature may reach 100°C. In industrial applications the ambient temperature is generally less than 50°C, and NEMA ratings for electrical insulation assume an ambient temperature of 40°C. In aerospace applications the ambient temperature may not be as important as the temperature of coolant provided by the airframe (oil or fuel); this may be as high as 200°F.

The increase in winding temperature increases the resistivity of the windings: a 50°C rise by 20%, and a 135° rise by 53%, increasing the I^2R losses by the same amount if the current remains the same. The increase in resistance is used in test procedures to determine the actual temperature rise of the winding, but this is obviously an average temperature; hot-spot temperatures can be 10-20° higher.

8.4 Heat Removal. Although there are many textbooks on heat transfer, the basic principles of heat transfer are summarized in a form that is useful for the motor design engineer [70]. [2] The three means of heat removal from a motor are *conduction*, *radiation*, and *convection*. Usually the most important is convection of air, or liquids, vapours, or oil mist. But if the motor is flange-mounted there may be appreciable conduction and consequent heating of the motor mounting. Radiation is generally small but *not negligible*, especially if the surface is enamelled, painted or lacquered black.

Conduction

The conduction equation for a block of thickness t and area A is

---

[2] Some of these formulas are adapted from Gary Skibinski's presentation to the *Brushless DC Drives* course at the Center for Continuing Education, University of Wisconsin, 1988, organized by the author.

$$Q = k A \frac{dT}{dx} \approx k A \frac{\Delta T}{t} \quad \text{W} \qquad [12]$$

where ΔT is the temperature difference across the block. The coefficient k is the *thermal conductivity*. This equation actually defines the thermal conductivity, with units (W/in^2) per (°C/in) or W/°C/in. [Sometimes this is written as W/°C-in. The SI unit is W/°C-m]. The thermal conductivity is a property of the material of which the block is made, and usually it is a function of temperature. The reader will be well aware that metals have high thermal conductivities, especially those which are also good electrical conductors. On the other hand, electrical insulating materials and most fluids have low thermal conductivities.

As an example, consider the flow of heat *along* a coil-side in the example motor in Chapter 10. The cross-section area of one coil-side is $A = 0.0794$ in^2, the stack length is 1.85 in, and the RMS current-density is 4,567 A/in^2, which produces heat at the rate of $J^2\rho = 4567^2 \times 0.678 \times 10^{-6} = 14.1$ W/in^3. In one coil-side the I^2R loss is therefore $14.1 \times 0.0794 \times 1.85 = 2.1$ W. To take the most pessimistic estimate, assume that all of this heat is generated at the mid-point of the coil-side, half-way along the stack. Then the temperature gradient along the coil-side is given by equation (12) as

$$\frac{dT}{dx} = \frac{Q}{kA} = \frac{2.1}{9.8 \times 0.0794} = 2.7 \text{ °C/in.} \qquad [13]$$

Since the heat can flow in both directions, the temperature-gradient is only half this value, and the temperature rise between the ends of the stack and the centre is therefore $2.7/2 \times 1.85/2 = 1.2$ °C, which is negligible. A more thorough analysis would have to consider the full *diffusion equation* along the length of the coil-side, but this quick calculation reveals that such sophistication is not needed in the example considered.

As a further example, assume that the example motor in Chapter 10 has a frame thickness of 0.3 in. Then the frame cross-section in the transverse plane is approximately $0.3 \times \pi \times 3.7 = 3.5$ in^2. If it is made of aluminium alloy with $k = 3.8$ W/°C-in, then for a temperature-gradient of 3 °C/in along its length the frame can conduct $3.5 \times 3.8 \times 3 = 40$ W, with a temperature difference of at most $3 \times 3 = 9°C$ along the motor length of about 3 in. This is about 15% of the motor output power. Therefore, if the motor efficiency is 80-85%, conduction to the flange could remove all of the motor losses. Clearly a flange mounting to a bulky piece of machinery or a "cold plate" can produce a significant contribution to the cooling, and this may make a large difference to the rating of the motor.

Contact resistance

Equation (12) can be used to define *thermal resistance* as the ratio of temperature difference ΔT to heat flow rate Q : the symbol usually used for thermal resistance is Θ, with units °C/W. Thus

$$\Theta = \frac{\Delta T}{Q} = \frac{t}{kA} \quad °C/W. \qquad [14]$$

The thermal resistance can be used to model the conduction through a region or interface where the individual values of k, A, and t are difficult to determine. The contact resistance between two surfaces bolted together is usually treated in this way. The temperature drop across a thermal resistance is given by equation (14) as $\Delta T = Q\Theta$. For example, if the contact resistance between the flange and the mounting plate in the previous example is 1°C/W, then with 40W flowing though it the temperature difference across the interface would be 40°C.

The contact resistance between metallic surfaces bolted tightly together depends on the surface finish. For a 120μin milled finish the heat transfer coefficient can be estimated as 0.7W/in^2/°C, corresponding to a thermal resistance of about 1.4°C/W for an area of 1in^2. A lapped surface (5μin) might have a value of half this. Thermal grease can be used to improve the heat transfer and lower the contact resistance by a factor of approximately 2 by replacing the voids which otherwise would be filled with air. The problem with thermal grease is that it tends to migrate. An alternative is to use a gasket which may be made of aluminium or copper foil, or special matrix materials impregnated with graphite or silicone.

In electric machines, contact resistances are a major source of uncertainty in thermal calculations, not only where the motor is externally bolted to something else, but also internally, wherever heat is required to flow across an interface. The most important interfaces are those between the coil conductors and the slot insulation, between the insulation and the lamination stack, and between the lamination stack and the frame. It may be impossible to calculate them, in which case test data is essential. The thermal contact resistances are usually the main impedances to the flow of heat from the interior of the motor to the outside, and consequently they make the largest contribution to the temperature rise of the hottest internal parts.

For this reason it is common practice in servo motors to encapsulate the windings in a resin of high thermal conductivity. Usually the thermal conductivity of the encapsulation is much less than that of a metal, but vastly better than that of air. In the most highly rated machines (aerospace machines), direct cooling of the conductors is often necessary because the overall thermal resistance between the windings and the outside world is simply too high to permit an acceptable power density.

The switched reluctance motor stands to gain appreciably from encapsulation of its windings because the windings tend to be large in cross-section, like the field coils of DC motors. The thermal diffusion length through such windings is much longer than in conventional brushless DC motors or induction motors with many small slots, and correspondingly the winding temperature-rise tends to be higher.

Radiation

Radiation is described by the *Stefan-Boltzmann* equation

$$\frac{Q}{A} = e\,\sigma\,(T_1{}^4 - T_2{}^4) \quad \text{W/in}^2 \qquad [15]$$

where σ is the *Stefan-Boltzmann* constant, 5.67×10^{-8} W/m^2/K^4 or 3.66×10^{-11} W/in^2/K^4 for a *black body*. A black body is a perfect radiator (no reflection). Real surfaces are imperfect radiators and their effectiveness relative to that of a black body is called the *emissivity e*. A black lacquered surface can achieve an emissivity as high as 0.98, but a more practical rule of thumb is to take 0.9 for black-painted or lacquered surfaces.

The temperature T_1 is the *absolute* temperature of the radiating surface, and T_2 is the absolute temperature of the surroundings radiating back to the motor.[3] For example, a surface with an emissivity of 0.9 that is 50°C above the surroundings at 50°C, has a net heat transfer rate of

$$0.9 \times 3.33 \times 10^{-11} \times ((50+50+273)^4 - (50+273)^4) \quad [16]$$

which is 0.28 W/in^2. A surface 30°C above the surroundings at 20°C has a rate of 0.12W/in^2 - quite a useful component of the heat-removal capability of the frame.

Convection

The rate at which heat is removed by convection is governed by *Newton's Law*:

$$\frac{Q}{A} = h \, \Delta T \quad \text{W/in}^2 \quad [17]$$

where ΔT is the temperature difference between the cooling medium and the surface being cooled, and h is the *heat-transfer coefficient*. The units of h are W/in^2/°C [or W/m^2/°C]. The value of h depends on the viscosity, thermal conductivity, specific heat, and other properties of the coolant, and also on its velocity. In *natural convection* the flow of coolant is not assisted by fans, blowers, pumps etc. In *forced convection* the flow is assisted by one of these external means.

Natural convection

The heat transfer coefficient for natural convection around a horizontally-mounted unfinned cylindrical motor can be roughly estimated as

$$h \approx 2.14 \times 10^{-3} \left(\frac{\Delta T}{D}\right)^{1/4} \quad \text{W/in}^2\text{/°C} \quad [18]$$

For example, for an unfinned cylinder of diameter 4in and a temperature rise of 50°C, the natural-convection heat-transfer

---

[3]The absolute temperature in degrees Kelvin (K) is the temperature in °C plus 273.

coefficient is calculated as 0.0040 W/in^2/°C. For a ΔT of 30°C the heat transfer rate is then 0.12W/in^2. As a first approximation this value can be applied to the whole surface including the ends, but if the motor is flange-mounted then only one end is available for convective cooling. For the example motor in Chapter 10, the surface area is approximately 60in^2 (excluding one end for flange-mounting), so the natural convection is approximately 7.2W.

Forced convection

Forced convection, with "air-over" cooling from a shaft-mounted or external fan, increases the heat-transfer coefficient by as much as 5-6 times, depending on the air velocity. The increase in heat-transfer coefficient is approximately proportional to the square-root of the air velocity. An approximate formula for the forced-convection heat-transfer coefficient is

$$ h \approx 11.2 \times 10^{-4} \sqrt{\frac{V}{L}} \quad \text{W/in}^2/°C \qquad [19] $$

where V is the air velocity [ft/min] and L is the frame length [in] (assumed parallel to the direction of airflow). For a motor of length 3.7in, if the air velocity is 800ft/min, this formula predicts $h = 0.0165$W/in^2/°C. This is 4 times higher than for natural convection. For the example motor in Chapter 10, with a cylindrical surface area of approximately 48in^2 (no fins), the convection rate for a ΔT of 30°C is 24W.[4]

The air velocity V is the *actual* air velocity, not the so-called "no-load" value. The no-load flow through a fan is usually specified in cubic feet/min (CFM), and the no-load velocity is given by the no-load CFM divided by the fan inlet area (in ft^2). The actual air velocity is determined from the intersection of the curve of static pressure vs. flow rate for the fan, and the pressure/flow curve for the air path over the motor. This calculation requires the use of fluid-dynamics, but a rough guide is to take V as one-half the no-load value:

---

[4]If the surface is finned, the appropriate area is the total surface area, which may be increased by a factor of 6 or more. This applies only if the airflow is along the fins.

$$V \approx \frac{\text{No-load CFM}}{2 \times \text{Fan inlet area}} \quad \text{ft/min} \qquad [20]$$

The size of fan required can be roughly estimated from the formula

$$\text{CFM} \approx \frac{1.76 \, Q}{\Delta T_{air}} \qquad [21]$$

where Q is the total rate of heat removal and ΔT_{air} is the temperature-rise of the air passing over the motor. Normally ΔT_{air} should be limited to about 15°C for an ambient temperature of 50°C. For example, if 100 W is to be removed with a ΔT_{air} of 15°C, the *actual* CFM of the fan must be 11.7 CFM. To allow for static pressure drop, a no-load fan of 25-30 CFM should be considered.

Some rules of thumb for "calibration"

Holman [70] gives an interesting example of a water-immersed wire 1m long, 1mm diameter, in which a power loss of 22W (0.56 W per inch length) is sufficient to boil the water at the wire surface. The wire surface temperature is 114°C and the heat transfer coefficient (see below) is 5000W/m^2/°C or 3.23W/in^2/°C. The heat flow out of the wire surface is 45W/in^2 (0.07W/mm^2) and the current-density in the wire is approximately 35A/mm^2.

In normal motors, the rate at which heat can be abstracted is *nothing like as high as this*. Correspondingly, current-densities as high as 35A/mm^2 are achievable only for very short bursts. This current density is sufficient to fuse a copper wire in free air.

The maximum rate at which heat can be removed from a surface by natural convection and radiation (with 40°C rise) is only about 0.5W/in^2. With forced air convection the rate increases to about 2W/in^2, and with direct liquid cooling about 4W/in^2. Motors that generate more heat than can be removed at these rates have to absorb the heat internally in their *thermal mass*, which is an acceptable way of increasing the output power for a short time.

These rates limit the heat generated *per unit volume* to about 0.2W/in^3 for natural convection, 5W/in^3 for metallic conduction, 7W/in^3 for forced-air convection, and 10W/in^3 for direct liquid cooling.

If rated torque is required at very low speed, a *shaft-mounted* fan may not provide enough coolant flow to keep the motor cool. DC motors often have separate AC-driven fans, because they have to work for prolonged periods at low speed with high torque. Since most of the heat in a DC motor is generated on the rotor (in the armature windings and the commutator), a good *internal* airflow is essential for cooling. In DC motors the external fan is usually mounted to one side of the motor, where it is easily accessible, and does not increase the overall length. A similar problem arises with AC induction motors, especially vector-controlled motors. A common practice is to mount the fan in line with the motor at the non-drive end, and arrange it to blow air over the outside of the finned frame. The fan may increase the overall length by as much as 60%. Brushless motors (including the switched reluctance motor) do not have this problem to the same degree, because most of the heat at low speed is generated in the stator windings, where all three forms of cooling (conduction, radiation and convection) are more effective. Even so, external fans are used on some highly-rated spindle motors.

Approximate heat transfer audit for the example motor in Chapter 10.

The following table summarizes the heat-transfer rates for the example motor in Chapter 10, using the approximate values calculated above for a case temperature rise of 30°C above a 20°C ambient.

The total heat transfer of 57.2W is quite close to the total losses of 53W (leaving a 4W allowance for windage loss, which was not calculated in the example), suggesting that the initial estimate of 30°C case temperature-rise is quite reasonable. However, it should be noted that without forced convection the total heat removal would be only 40.4W (67% of the losses), so the case temperature would rise if full-load operation were sustained.

Also note that conduction to the flange mounting accounts for nearly half the total heat removal. If this were not permitted (for example, if the user wished the motor to cause *no* heating of his equipment), the rating would be reduced by the order of 50% and it would be difficult to increase it by other means short of halving the internal motor losses.

Heat transfer audit for example motor in Chapter 10		
Frame size	4in dia × 3.7in Length	
Surface finish	Black	
Total losses (incl. 4W windage)	57W	100%
Natural convection	7.2W	12.6%
Forced convection	24W	42.0%
Conduction to flange mounting	26W	45.5%
Radiation assuming 30°C case temp. rise above 20°C ambient	7.2W	12.6%
Total heat removal (forced convection)	57.2W	100%

8.5 Internal temperature distribution. The heat-removal calculation only tells us whether a steady-state case temperature can be achieved for a given value of total losses, and of course it helps us decide the method of cooling. But it tells nothing about the internal temperatures, and gives no guidance whatsoever on the values of current-density, flux-density, and frequency that can be used.

To answer these questions it is necessary to use a *thermal equivalent circuit* of the interior of the motor, Fig. 8.2. The thermal equivalent circuit is an *analogy* of an electric circuit, in which heat is generated by "current sources" and temperature is analogous to voltage. The rate of generation of heat in a source is measured in Watts. The heat flow rate, which is also measured in Watts, is analogous to current. Resistance is measured in °C/W. The I^2R losses, core losses, and windage & friction losses are represented by individual current sources, and the thermal resistances of the laminations, insulation, frame, etc. are represented as resistances. In the simplest possible model, all the losses are represented together as one total source, i.e. the individual sources are taken as being in parallel.

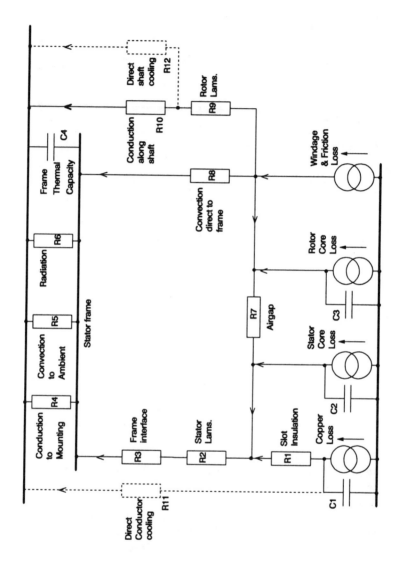

Fig. 8.2   Thermal equivalent circuit

The thermal equivalent circuit is actually a simple model of the *heat conduction equation* or diffusion equation. The complete form of this equation is

$$\nabla^2 T + \frac{1}{k}\frac{\partial q}{\partial t} = \frac{1}{\alpha}\frac{\partial T}{\partial t} \qquad [22]$$

where

$$\nabla^2 T = \frac{\partial^2 T}{\partial x^2} + \frac{\partial^2 T}{\partial y^2} + \frac{\partial^2 T}{\partial z^2} \qquad [23]$$

and

$$\alpha = \frac{k}{\rho c} \quad \text{m}^2/\text{s} \qquad [24]$$

is the *diffusivity* in SI units. (English units are in^2/s). In SI units, k is the thermal conductivity in W/m°C; c is the specific heat in kJ/kg°C, and ρ is the density in kg/m^3. In a structure as complex as an electric motor the heat conduction equation is a complex boundary-value problem that is best solved by computer-based numerical mehthods such as the finite-element method. However, for most practical purposes a simple *lumped-parameter* model can be constructed relatively easily. It should ideally take into account the anisotropy effects: for example, the effective thermal conductivity through a lamination stack is lower in the axial direction than in the radial direction.

A thermal equivalent circuit for the switched reluctance motor is shown in Fig. 8.2. The circuit includes provision for direct cooling of the winding conductors, Θ_{dcc} (R11) and for direct cooling of the rotor shaft Θ_{drc} (R12). It also includes the thermal mass or *capacity* of the winding C_w (C1), and the thermal capacity of the rotor and stator laminations, C_{rl} (C3) and C_{sl} (C2). The other internal thermal resistances are essentially self-explanatory.

The heat removal routes by conduction, radiation, and convection are also represented by thermal resistances. For convection the appropriate resistance Θ_v is given by

$$\Theta_v = \frac{1}{hA} \quad °\text{C/W} \qquad [25]$$

where A is the appropriate surface area for convective heat-transfer. If h is a function of temperature-difference, the equivalent circuit becomes non-linear and requires an iterative solution (e.g. Newton-Raphson method). For radiation the equivalent thermal resistance is the ratio of the temperature difference $T_1 - T_2$ to the radiation heat exchange rate Q in equation (15). Clearly this is non-linear. However, the non-linearity is often neglected and a fixed value of θ_r is calculated assuming that the final temperature of the case is known.

Current Density

The current-density cannot be directly related to the temperature rise of the winding by a simple general equation, because it depends on the shape of the conductors. For example, $1in^3$ of copper can be made into a stubby cylinder of 1in diameter and 1.27in length, or a long wire of 0.5mm diameter and 3,286in length. If only the cylindrical surface area is available for cooling, the short cylinder has a surface area of $4in^2$ while the long wire has a surface area of $203in^2$. The loss density in W/in^3 in copper conductor is $J^2\rho$ where J is in A/in^2 and ρ is in ρ-in. At 20C the resistivity of copper is approximately 0.66 micrOhm-in, but at 200C it is about 1.16 micrOhm-in. If 1W can be dissipated from every in^2 of surface, this suggests that in the short stubby cylinder the permissible current density is $1,857A/in^2$ ($2.88A/mm^2$) and in the long wire, $13,240A/in^2$ ($20.5A/mm^2$).

With this reservation, it is still possible to quote typical values of current densities used in motors cooled by different methods [71]:

Current density - typical values		
Cooling method	A/in^2	A/mm^2
TENV	3000 - 3500	4.7 - 5.4
Air over; fan-cooled	5000 - 7000	7.8 - 10.9
External blower; through-cooled	9000 - 10000	14.0 - 15.5
Liquid-cooled	15000 - 20000	23.3 - 31.0

Selected material properties				
Material	ρ Ω-m $\times 10^{-8}$ 20°C	k (W/in²) per (°C/in)	Sp. Heat kJ/kg/°C	Density kg/m³
Copper (pure)	1.7241	9.80	0.383	8954
Aluminium (pure)	2.7	5.18	0.896	2707
Aluminium (alloy)		3.8		
Silicon steel		0.79		
0.5% carbon steel		1.37	0.465	7833
Iron (pure)		1.85	0.452	7897
Insulation		0.035		
Ferrite magnets		0.012		
Water (20°C)		0.0153	4.18	997.4
Freon		0.0019	0.966	1330
Ethylene Glycol		0.0063	2.38	1117
Engine oil		0.0037	1.88	888

Temperature rise by resistance and Insulation Class			
Motor type	Class **B**	Class **F**	Class **H**
1.15 Service Factor	90	115	
1.00 Service Factor	85	110	
TEFC	80	105	125
TENV	85	110	135

(NEMA Standard MG-1), °C. Assumes 40°C ambient temperature.

Selected emissivities	
Material	Emissivity
Polished aluminium	0.04
Polished copper	0.025
Mild steel	0.2-0.3
Grey iron	0.3
Stainless steel	0.5-0.6
Black lacquer	0.9-0.95
Aluminium paint	0.5

9 Applications

9.1 General. Electric motors and drives are almost always sold into niche markets, their design specialized to the application. Some "niches" are, of course, very large. Like other motors and controls, the switched reluctance motor has a unique set of characteristics that make it suitable for some applications and unsuitable for others.

Articles often list the advantages and disadvantages of the switched reluctance motor but fail to follow through with the *logic* of using it for a particular application. With hindsight it is easy to see what that logic is. This chapter begins with a summary of first impressions of the switched reluctance motor, followed by a more critical analysis. Then a number of example applications and products are reviewed.

9.2 Motor: first impressions. The most striking feature is the simplicity of mechanical construction, which suggests the following positive and negative impressions:[1]

	Motor - positive impressions		Motor - Negative impressions
P1	Low manufacturing cost	N1	High scrap in punching
P2	Low material cost	N2	Small shaft diameter
P3	Minimal temperature effects	N3	Small airgap
P4	High-speed operation possible	N4	Large coil cross-section may lead to hot-spots
P5	Low inertia	N5	Shaft position sensor needed
P6	Ease of repair	N6	Doubly-salient structure will cause noise, torque ripple
P7	Short end-turns, with no crossovers	N7	Apparently high windage loss at high speed
P8	Low rotor losses	N8	Long, two-pole flux-path
P9	Fault tolerant	N9	Cannot start direct on-line

---

[1]The positive and negative impressions are labelled P1,P2, N1,N2... for reference in the following analysis.

9.3 Controller: first impressions. The controller may represent more than half the cost of the whole drive. First impressions here can be even less reliable than those formed about the motor. The most obvious positive ones are as follows:

	Controller - positive impressions		Controller - Negative impressions
P9	The number of transistors is the same as, or less than, the number required for an AC drive with the same number of phases	N9	The circuit cannot use phaseleg modules developed for AC inverters
P10	The use of one half-bridge per phase winding provides protection against shoot-through failures	N10	Both ends of every phase winding appear to need a connection
P11	There is a high degree of independence between phases	N11	The non-uniform torque may lead to a high filter-capacitor requirement
P12	There is a wide variety of circuit topologies for different applications.	N12	The non-uniform torque/ampere may limit the servo bandwidth
P13	The control is well suited for digital implementation with linear current regulation	N13	High di/dt is a possibility, mandating the use of short, low-inductance cables
P14	"Reluctance motors have a wide speed range at constant power"	N14	"Reluctance motors have a high kVA requirement"
P15	No open-circuit voltage	N15	Commutation frequency is higher than in AC motor with the same rotor pole number
P16	No sustained short-circuit current	N16	Lack of short-circuit current may limit applications as a generator; no dynamic braking

9.3 Critique of motor characteristics.

P1 Low manufacturing cost. The number of processes involved in manufacturing switched reluctance motors is comparable to the number required for induction motors. There is a saving in that no rotor casting is required, but on the other hand there is a need for some means of compressing the rotor laminations in the axial direction. This can be as simple as a nut and washer (e.g. a Belleville washer); there will be a small additional cost associated with screw-threading the shaft. Because of the simplicity of the laminations and the windings, the switched reluctance motor may require less capital investment in winding machinery and rotor-casting machinery. In small numbers it can be manufactured by essentially routine production-engineering, and this may make it suitable for integration in certain system-level products.

The absence of magnets means that all the manufacturing costs associated with magnets are saved: testing purchased magnets, magnetization, bonding and/or the assembly of retaining rings. The problems of handling magnetized magnets and rotors in the factory are also avoided.

Some unpublished analyses for particular applications have shown that the manufacturing costs of the switched reluctance motor can be as low as 60% of those of the DC or AC motor which it is intended to replace. The controller, on the other hand, does not have any significant cost-saving features compared with those of the AC inverter.

P2 Low material cost. The low material cost stems from the absence of permanent magnets and aluminium (since there is no rotor cage winding). Most of the materials are readily available - electrical sheet steel, magnet wire, and engineered parts such as shafts, bearings, frame components, etc.

Because of the higher switching frequency for a given speed, the switched reluctance motor may require the use of thinner laminations and a higher grade of core steel, as compared with a *standard-efficiency* induction motor; but the switched reluctance motor should be compared against the *high*-efficiency induction motor, which also requires a higher grade of core steel, thinner

Page 137

laminations, and a longer stack. The switched reluctance motor tends to be copper-loss intensive, and requires a slightly larger volume of copper than the high-efficiency induction motor to achieve the same efficiency. The volume of iron is likely to be lower, but on the other hand the scrap is greater.

P3   Minimal temperature effects. Obviously the switched reluctance motor is free from the problems of partial or complete demagnetization that can occur in PM motors. It is true, of course, that permanent magnet materials (e.g. Sm_2Co_{17}) can be used up to 200-250°C, but they are extremely expensive. The induction motor has its own temperature-related problems. Rotor losses, which may be difficult to remove, cause increased slip and potential thermal instability. Variations of the rotor losses cause variations in rotor temperature that can impair the mechanical integrity by inducing rotor-bar fractures. The variation of rotor resistance also changes the torque per ampere, making for increased complexity in the controllers (which normally require shaft-position encoders). AC and DC commutator motors are also subject to limits on commutator temperature. Temperature effects in the switched reluctance motor are essentially limited to variations of stator resistance. If the temperature rise is too high there may be accelerated aging of the stator insulation and problems with bearing lubrication; but these problems are no worse than in the other motor types.

Unfortunately the switched reluctance motor tends to have a high electric loading that can lead to a high temperature rise. This does not restrict its ability to match or exceed the power density of the induction motor, but it does limit the overload rating and the ability to convert all the available electromagnetic energy that appears to be available in the flux-linkage/current diagram.

P4   High-speed operation possible. The maximum speed is limited by five main factors :

- (**a**) the core losses and windage losses;
- (**b**) the strength of the rotor steel;
- (**c**) the shaft dynamics;
- (**d**) the bearings; and
- (**e**) the volt-ampere rating of the controller.

(a) The core losses increase rapidly with frequency. In particular, in linear theory the eddy-current component increases in proportion to B^2f^2, suggesting that for constant loss-density the flux-density must be decreased inversely with speed such that the product Bf remains constant. Incidentally, this also implies that it should be possible to maintain a constant-power characteristic (with torque × speed = constant), other things being equal. At very high frequencies, however, the eddy-current losses increase faster than f^2. One reason is that the laminations are no longer thin enough to keep the eddy-currents in the "resistance-limited" condition. In principle the problem can be solved by decreasing the lamination thickness, but in practice there is a lower limit set by the increase in handling costs, the fragility of the individual laminations, and the reduction in the stacking factor. The thinnest laminations in common use are usually 0.006in for cobalt-iron and 0.35mm (0.014in) for silicon steel, but occasionally laminations as thin as 0.004in or even 0.002in are used. 6-mil laminations of cobalt-iron are used in some aircraft generators which operate at 400Hz, but others use silicon steel. Much higher frequencies are experienced in high-speed switched-reluctance motors but, in mitigation, the volume of iron tends to be quite low and the highest flux-densities tend to be localised in relatively small volumes.

(b) At extremely high speeds the tensile stress (hoop stress) in the rotor yoke can produce sufficient strain to cause the rotor laminations to "lift" off the shaft. It is desirable to assemble the rotor of a high-speed machine with an interference fit such that, even at the highest speed, the centrifugal loading does not reduce the interference to zero. A crude estimate of the effect of centrifugal loading can be made as follows. Consider the rotor of the example motor in Chapter 10, Fig. 10.3. The rotor is shown again in Fig. 9.1, with the four poles and adjacent sections of rotor yoke sectioned off. The weight of one of these four sections is approximately

$$w \; \approx \; \rho L_{stk} [t_r d_r + \frac{\pi}{4}(r_0{}^2 - r_{sh}{}^2)] \tag{1}$$

where the symbols are defined in Table 10.5 and $r_{sh} = D_{sh}/2$, $r_0 = r_{sh} + y_r$. Substitution of the dimensions from Table 10.5 gives $w = 0.0944$kg. Taking the mean radius as $r_{mean} \approx r_0 = 0.683$in (17.35mm), the centrifugal load on this section is $r_{mean}w\omega^2$. At 50,000rpm this has a value of 44,898N, which must be borne by the tensile hoop

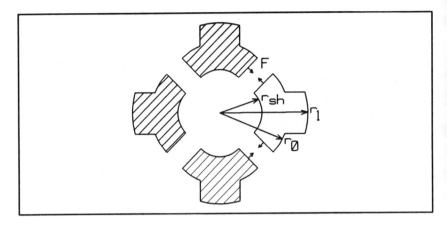

Fig. 9.1   Hoop stress in high-speed rotor

forces labelled F in Fig. 9.1. Taking into account the directions of these forces, each force F across the yoke section is 44,898/(2 cos 45°) = 31,747N or 7,128lbf. The corresponding stress is this force divided by the yoke section in the r-z plane, i.e. (0.340 × 1.85) = 0.629in^2. Therefore the hoop stress is 7,134/0.629 = 11,332lbf/in^2. This is quite small compared with the yield strength, and with a Young's modulus of 30 × 10^6 the strain would be only of the order of 0.04%. This gives a rough initial idea of the interference fit required, i.e. 0.04% of the shaft diameter or about 0.26 mils.

If all the rotor dimensions were doubled, the hoop stress would increase by a factor of 4 to some 45,000lbf/in^2, which is comparable with the yield strength of normal silicon steels. At this speed and scale it becomes necessary to select the rotor steel for its mechanical properties as much as for its electrical properties, and the heat-treatment must be carefully optimized also. Fortunately cobalt-iron, which is used almost exclusively in high-speed, highly-rated machines, has a higher yield strength than silicon steel when correctly heat-treated, although there is a tradeoff between mechanical strength and core losses in the annealing temperature.

P5 Low inertia. The low inertia of the switched reluctance motor obviously arises from the large chunks of material missing in the interpolar spaces of the rotor. The tendency is also to design with a low rotor/stator diameter ratio and a long length/diameter ratio.

Reference [67] describes a 10-hp switched reluctance motor with a T/J ratio of 2490rad/sec^2 compared with a high-efficiency induction motor having a ratio of only 910rad/sec^2, an advantage of 2.75:1. These are quite large motors. In smaller machines higher peak T/J ratios are possible, and figures as high as 80,000rad/sec^2 have been quoted for cylindrical-rotor machines. Of course, much higher figures (up to 10^{6}rad/sec^2) can be achieved with special ironless rotors but such machines typically do not have very high power capability.

P6 Ease of repair. The ease of repair stems from the simplicity of the stator windings. In some cases it is possible to remove and rewind the coil from one pole without disturbing its neighbours.

P7 Short end-turns. The advantage of having short end-turns is that little copper is wasted in the end-turns, the phase resistance and copper losses are reduced, and the active core length occupies a higher proportion of the overall frame length than in other types of motor, with the possible exception of brushless DC motors that are wound with one coil per tooth. The disadvantage of having short end-turns is that less surface area is available for cooling by shaft-mounted fans at the ends of the machine, and greater reliance must therefore be placed on the conduction of heat radially outwards through the frame. Short end-turns are also mechanically rugged, being less subject to movement and fretting. In the switched reluctance motor there are no cross-overs between coils of different phases, so that the possibility of a phase-to-phase fault is reduced.

P8 Low rotor losses. The rotor losses are certainly not zero, as has been sometimes stated, but they are certainly lower than in induction motors. The point should be made that rotor losses are not zero even in permanent-magnet rotors, especially if a retaining can is used to contain the magnets. In the switched reluctance motor the rotor core losses are often negligible at moderate speeds. Although the frequency of alternation is higher than in AC motors, the magnetic loading and the volume of steel are both low. At very high speeds the rotor losses can be a limiting factor and it may be difficult to achieve a satisfactory compromise between core loss density and mechanical strength. In cobalt-iron, for example, which is commonly used for high-speed machines in aerospace applications, there is a critical range of annealing temperature to achieve the best compromise between these properties.

P9   Fault tolerant.   The switched reluctance motor is "fault tolerant" in the sense that if an open-circuit or short-circuit fault occurs in one of the phase coils, the consequences are unlikely to be dangerous. In permanently excited machines and even induction motors, the rotating flux can induce extremely high current in shorted windings or shorted turns. In the PM machine there is no means to shut off the excitation, and in large induction motors the time constant associated with trapped rotor flux may be long enough for the short-circuit current to do serious damage. This possibility is eliminated in the switched reluctance motor.

N1   High scrap in punching.   A glance at the typical switched reluctance motor lamination set shows this to be broadly true. However, the spaces punched out of the rotor do not have to be filled with cast aluminium, as they do in the induction motor, and they also help to promote good cooling. The scrap content in the stator can be minimized by using a hexagonal or square outside profile, as shown in Figs. 3.5, 3.6, and 3.16.

N2   Small shaft diameter.   The rotor/stator diameter ratio is typically smaller than in AC motors, and smaller still when compared with the ratios used in DC motors. This is a result of the tendency to design for maximum inductance ratio while maintaining sufficient yoke sections to keep the flux-densities low and to maintain a sufficiently stiff stator structure against ovalizing forces. The disadvantage of a small shaft diameter is that it leads to a low critical speed, and increases the possibility of lateral deflection that could cause the rotor to touch the stator, under unbalanced magnetic forces.

N3   Small airgap.   There is no doubt that a small airgap is desirable to achieve the highest inductance ratio and thereby maximize the torque per unit volume and the efficiency, and to minimize the converter volt-amperes. The airgap need be no smaller than that of the induction motor. However, there is some evidence that the switched reluctance motor is more sensitive to *variations* in the airgap than the induction motor, in that any variation may affect the balance between phases and may have an adverse effect on the noise level. This is a disadvantage, because a tight tolerance on concentricity may make the machine more expensive. The switched reluctance motor cannot tolerate the large airgaps used in certain

Page 142

types of PM motor. However, PM motors with retaining rings also require small mechanical airgaps. The choice of airgap length for the switched reluctance motor is discussed in §10.6.

N4 Large coil cross-section may lead to hot-spots. This disadvantage may limit the maximum permissible current density, although it is offset by the shortness of the end-turns and the consequent reduction in total copper loss. See Fig. 3.30.

N5 Shaft position sensor needed. This has been discussed in Chapter 7, where it is shown that sensorless operation is feasible in certain circumstances. AC motors used in field-oriented control also require shaft encoders, and they typically use 1000-line optical encoders. The switched reluctance products currently sold in industrial applications are not impaired by having shaft position sensors, as many of them are sold in place of DC motors which require frequent maintenance to commutator and brushgear.

N6 Doubly salient structure causes noise and torque ripple. Noise level can be reduced by using a stiffer structure, decreasing the current ripple, and adjusting the firing angles; but the likelihood of achieving very low noise levels at the same power density as a permanent magnet motor or a carefully designed induction motor seems remote. As discussed in Chapter 7, elimination of torque ripple is achievable only at the expense of considerable control complexity and very precise shaft position feedback.

N7 High windage loss at high speed. The scalloped shape of the rotor is likely to induce additional pumping and turbulence resulting in high windage loss at high speed, especially at extreme speeds where it is not feasible to fill the rotor slots. At lower speeds the rotor slots can be bridged or filled, but this incurs additional expense and compromises the simplicity and ruggedness; fortunately it is rarely necessary.

N8 Long, two-pole flux-path. As discussed in Chapter 3, there are designs of switched reluctance motor that have short flux-paths but the penalty is usually a higher number of poles and/or phases with a correspondingly smaller stroke angle and higher switching frequency for a given speed.

N9 Cannot start direct on-line. Although it is the case that induction and DC motors can start direct on-line, it is easily forgotten that when they do so, they draw many times rated current and cause a severe (though temporary) disruption to the local power supply. This actually makes it difficult for protective equipment to discriminate between fault current and starting current unless the protection is disabled during starting. One of the most widely used items of power-electronic equipment in factories is the *soft-starter*, which is designed to limit the starting current. Unfortunately, soft starters prolong the acceleration time because they reduce the RMS voltage applied to the motor without altering the frequency. They also introduce severe harmonic distortion during the run-up period. The switched reluctance motor can always start with at least rated torque from standstill without drawing more than rated current from the supply. It is equivalent in this respect to the inverter-fed induction motor. The only advantage of the inverter-fed induction motor is that if the inverter fails, in theory the induction motor could be started by reconnecting it direct to the line. But if the load can operate satisfactorily through the uncontrolled acceleration, and subsequently at constant speed, why was the inverter fitted in the first place?

9.4 Critique of controller characteristics.

P9 The number of transistors is the same as, or less than, the number required for an AC motor with the same number of phases. This is discussed in Chapter 6. In low-cost mass-produced applications the switched reluctance motor clearly has an advantage. In higher-power 3-phase drives the circuit of Fig. 6.1 is likely to be the best choice, with no advantage over the 3-phase AC drive. But four-phase switched reluctance drives can be made with 4 or 6 transistors, as shown in Figs. 6.5 and 6.7.

P10 The use of one half-bridge per phase provides shoot-through protection. Although modern phaseleg modules for AC drives are relatively immune against shoot-through, it remains a possibility in the AC type inverter bridge. The only way the shoot-through failure can occur in the switched reluctance motor controller is for the phase winding itself to be short-circuited. If that happens, of course, simultaneous turn-on of the power switches into the faulted winding will render them liable to destruction unless they are fully protected against sudden short-circuit. In AC drives, a full phase-phase fault would have the same effect.

P11 There is a high degree of independence between the phases. This is true for circuits which use two switches per phase, with no sharing of transistors between phases. It means that if one phase is out of action, the others can carry on unaffected and the power output is reduced in proportion. Because of the nature of the control, the loss of controllability associated with the loss of one phase is relatively small except at low speeds. In generator applications, obviously there is an advantage in being able to operate with one phase faulted, but also the phases can be connected to isolated controllers to supply several different channels, which may or may not be combined in the power electronics.

P12 There is a wide variety of circuit topologies for different applications. This variety stems from the ability to operate with fewer than 2 switches per phase, which is not generally feasible with AC motors. Additional variety comes from the relatively free choice of phase number, which is due to the fact that the rotor and stator pole numbers do not have to be equal as they are in AC motors.

P13 The control is well suited for digital implementation. Commutation is a timing function similar to that of the brushless DC motor. The optimization of the performance of the switched reluctance motor requires firing angle variation according to speed and load torque, and this is best implemented digitally. In fact, all functions in the control of the switched reluctance motor are readily implemented using standard electronic components, without the complexities of vector control required by induction motors.

P14 Reluctance motors have a wide speed range at constant power. The "wide" speed range of the switched reluctance motor is an impression formed from the series-type torque/speed characteristic and the ability to boost the torque at high speed by advancing the turn-on angle and increasing the dwell angle. However, there is little published evidence to show that the switched reluctance motor has any advantage. It is certainly not able to outperform the separately-excited DC motor [72] in terms of the field-weakening range at constant power. Moreover, difficulties in obtaining smooth operation at low speeds (below 100-300rpm) may introduce a limit on the useable speed *ratio*, regardless of whether torque or power is maintained constant.

P15 No open-circuit voltage. Because of the absence of permanent excitation there is no generated voltage on open-circuit. This may be an advantage in applications with freely over-running loads, or during fault conditions in safety-critical applications. In contrast, over-running PM brushless motors become generators at high speed and their output is rectified back on to the DC supply in an uncontrolled manner. In certain applications this necessitates the use of powerful overvoltage protection on the DC supply.

P16 No sustained short-circuit current. Because of the absence of permanent excitation there is no generated voltage to drive short-circuit current into a short-circuit fault that has occurred in the controller. However, a short-circuited phase is likely to have current induced due to the action of the other phases, and is also likely to affect their operation.

N9 The controller cannot use phaseleg modules developed for AC drives. This is a disadvantage in that the layout of the controller is less tidy when individual devices (transistors and separate diodes) are used, because there is more buswork and it is more difficult to achieve compactness. Probably the stray inductances and EMI problems are correspondingly increased compared with AC drives.

N10 Both ends of every phase winding appear to need a connection. Unlike the AC 3-phase motor which can be connected in star (wye) or delta, the switched reluctance motor generally has no common winding connections except in circuits such as Figs. 6.5 and 6.7. Accordingly the number of terminals and cables tends to be higher than in AC drives.

N11 The non-uniform torque may lead to a higher filter capacitor requirement. This is inevitable, and arises from the fact that a phase is fully fluxed from zero and then de-fluxed every stroke. The same is true in AC and brushless DC motors, but in PM motors the fluxing and de-fluxing is accomplished by the rotation of the magnet and no magnetic energy need be drawn from or returned to the DC supply.

N12 The nonlinear torque/ampere may limit the servo bandwidth. Not enough work has been done on the dynamics of switched reluctance motor controllers to understand fully the effects of the nonlinearities. However, the torque/ampere varies considerably with rotor position, implying that gains in the feedback loop should be limited or at least "scheduled" to compensate for this effect without risking instability or magnifying speed ripple.

N13 High *di/dt* is a possibility. The minimum inductance of the switched reluctance motor is very low, and in systems with a low supply voltage and/or a very wide speed range the inductance of supply cables can absorb supply voltage and deplete the performance of the motor. Furthermore, the high *di/dt* obtained with full DC voltage connected during chopping can induce voltages in neighboring circuits.

N14 Reluctance motors have a high kVA requirement. The kVA requirement of switches in switched reluctance motor controllers is broadly comparable to that of AC drives. This has been demonstrated in [28] and [16]. (See Chapter 10).

N15 Commutation frequency is higher than in an AC motor with the same rotor pole number. Although this is the case, the commutation frequency is not particularly important because the switching frequency in chopping is much higher and chopping is required at all lower speeds.

N16 Lack of short-circuit current may limit applications as a generator; no dynamic braking. In conventional AC systems, circuit-breakers are activated as a result of the detection of fault current many times normal current and sustained for one or more cycles. If the circuit-breakers were set to trip on the occurrence of shorter-duration fault currents, there would be more nuisance-tripping. It is part of the philosophy of AC power systems that the generators should be capable of delivering many times their rated current when they are short-circuited. The switched reluctance motor is connected to the power system through a power electronic converter which is dominated by the filter capacitor (See chapter 6). If it is used as a DC generator the voltage across the filter capacitor is liable to collapse on the occurrence of a short circuit. This may be an advantage in a DC power system, provided that there are independent means for identifying and isolating faults. The lack of short-circuit current capability (after the rapid discharge of the filter capacitor) appears to remove the need to interrupt large DC currents at high voltage. If the switched reluctance machine is used as an AC generator, the DC must be re-inverted electronically and there is liable to be a problem with conventional systems due to the lack of short-circuit current.

The lack of open-circuit voltage and short-circuit current also means that there is no dynamic braking when the phase windings are short-circuited or connected to a braking resistor. This type of braking is very effective with PM brushless motors and even with induction motors. There is some evidence that switched reluctance motors can be braked very rapidly by connecting a braking resistor in parallel with all of the commutating transistors while continuing to control the supply current with the chopping transistors [73].

9.5 Application examples.

The following pages show photographs of examples of switched reluctance motors developed for particular applications or market sectors, kindly supplied by the manufacturers or designers concerned. The photographs represent only a small fraction of the development work that has been done, but they represent a large fraction of the products currently in European production. Products made in the United States and Japan have been described in other publications, but a partial list of manufacturers and their switched reluctance products is included at the end of this chapter. This list also mentions non-manufacturing organizations who have significant research and development activity in the field of switched reluctance motors. The list is not exhaustive, and deliberately omits a number of manufacturing companies who may have research and development programmes but who have not announced products.

Fig. 9.2   20kVA, 15,000rpm prototype starter-generator rotor, courtesy Lucas Advanced Engineering Centre

Fig. 9.3   *OULTON* SL variable-speed SR drive; up to 45kW.
Courtesy Graseby Controls, UK

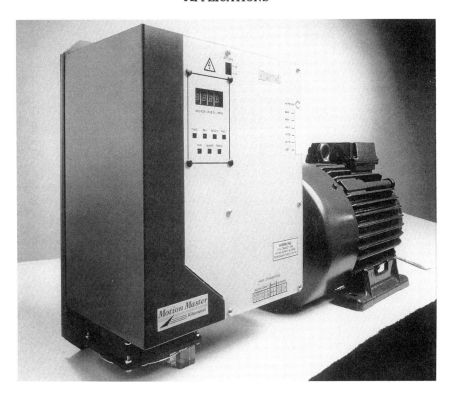

Fig. 9.4   Allenwest variable-speed SR drive; 7.5-22kW. Courtesy
             Allenwest, UK

Fig. 9.5   Battery-powered electric vehicle with SR traction drive developed by HIL, UK. Courtesy of HIL Ltd, UK

Fig 9.5.1   4 Phase SR adjustable RPM Drives .25 to 10KW Courtesy Magna Physics division of Tridelta Industries Hillsboro, Ohio.

Fig. 9.6   2kW, 4-quadrant high-performance pallet-truck drive. Courtesy SRDL, UK

Fig. 9.7   Selection of SR motors ranging from 100W at 5,000rpm
to 100kW at 1500rpm. Courtesy SRDL, UK

Fig. 9.8   Water-cooled 35kW SR motor for mining and other
            heavy-duty applications. Courtesy British Jeffrey
            Diamond, Wakefield, UK

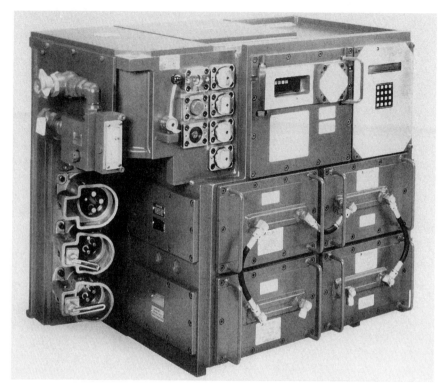

Fig. 9.9   Controller for water-cooled motor of Fig. 9.8. Courtesy
British Jeffrey Diamond, Wakefield, UK

Fig. 9.10   Food-processor drive comprising SR motor and single-
stage gear. The flat profile permits a complete re-design
of the food processor. Courtesy SRDL and Smallfry Ltd,
UK

Fig. 9.11   24V, 6,000rpm prototype SR motor showing laminations and one coil. Courtesy Lucas Advanced Engineering Centre, UK

Organization	Product
Magna Physics Corporation, Hillsboro, Ohio JR Hendershot	Switched reluctance motors and generators; CAD software; PM brushless DC motors; consultancy
Switched Reluctance Drives Ltd, Leeds, England Prof PJ Lawrenson	Design, development and licensing; prototype motors and controllers; applications embracing white goods, position servo, general-purpose industrial, mining, ultra-high speed, automotive; 37 employees; since 1980
SPEED Laboratory, Glasgow University Prof TJE Miller	Design consultancy; *PC-SRD*, CAD software; prototype motors and controllers; research
Virginia Polytechnic Institute Prof R Krishnan	Design consultancy; CAD software; motors & controllers; research
MIT Prof J Lang	Design consultancy; Research; prototype motors and controllers
Graseby Controls, Lowestoft, UK (Licensee of SRDL)	4-80kW industrial general-purpose SR variable-speed drives up to 45kW including EEx. Also: Eddy-current drives and AC motor controls
Allenwest, Prestwick, Scotland, UK (Licensee of SRDL)	General-purpose SR variable-speed drives 7.5-22kW *Motionmaster*
British Jeffrey Diamond (Dresser), Wakefield, UK (Licensee of SRDL)	4-quadrant water-cooled SR drives for mining and other very heavy-duty applications 35-200kW
Brother Manufacturing Co, Japan	Ultra high-speed spindle and servo motors up to 100,000rpm; 2hp
Warner Electric, USA	Small SR motors and stepper motors
Hewlett Packard, USA	HCTL 1100 control IC for VR and PM brushless motors, includes motion control profiling; Draftmaster plotter incorporates 2 SR servo drives
Semifusion Corp, Santa Clara, USA	SR controllers, fully user-programmable; 2-axis servo drives
National Semiconductor, Santa Clara, USA	LMB1008 controller (low performance); HPC46883 16-bit microcontroller (high performance)
Motornetics Corp, Santa Rosa, USA	Megatorque direct-drive double-airgap VR motors; 5rps, 230ft-lb

Page 159

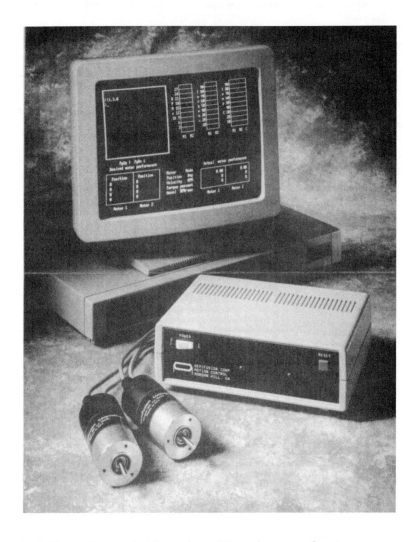

Fig 9.12　　Two axis (4) quadrant SR motion control system.
Courtesy Semifusion Corp, Morgan Hill, CA

10 EXAMPLE DESIGN AND ANALYSIS

10.1 Design process. A *specification* consists of *requirements* (e.g. torque, speed) and *constraints* (e.g. temperature-rise, supply voltage, dimensions). A *design* consists of *parameters* defining the motor and controller, and includes the specification of materials and manufacturing processes. Specifications sometimes include the *envelope dimensions* (overall dimensions); if they do not, one of the first tasks in design is to make an initial estimate of them, §10.4. Once the envelope dimensions are known, estimates of internal dimensions can be initiated by "ratio-ing" from the envelope dimensions, §10.6ff, starting with "standard" proportions.

After initial dimensioning there follows a process of refinement in which, by "chipping away" at the design parameters one-by-one, the designer gradually improves the performance and quality of his/her design. The refinement process is facilitated by appropriate computer software, by test data from prototypes, and above all by experience. Remarkable design improvements continue to be accumulated in this way, even for mature products like DC motors. The switched reluctance motor is only at the start of this process.

10.2 Specification. A specification should be *precise* and *complete*. Where possible the user and designer should negotiate the specification in the interests of getting the best overall design. It is unsatisfactory to design motors in isolation from their end-use, particularly with adjustable-speed drives. It is a mistake to *overspecify*, that is, to require things that are irrelevant or not necessary. A common example is to specify a motor bigger than the load requirement. On the other hand, the supplier should never be starved of information that may help him/her make a better design.

The most basic requirements in a specification are *torque, speed, load factor* and *supply voltage*. It is essential to specify whether the motor must be able to rotate in both directions or not; and whether braking or generating operation is required; this refers to the *quadrants* in which the motor is to operate, Fig. 10.1. Ambient temperature, variations in supply voltage, and special factors should also be included, as well as any standards or regulations which must be met.

The table overleaf specifies the example worked out in this chapter.

	Example specification	
	Requirement or Constraint	**Example**
Must be specified for all designs	Maximum torque Speed Load factor Supply voltage Forward/reverse Motoring/generating	10 lbf-in 2,000 rpm 50% 24 Yes No
Desirable to specify	Temperature rise Envelope dimensions Overload rating / time Compliance with standards	
Examples of other requirements and constraints	Type of enclosure Maximum noise level Maximum harmonic current Maximum EMI/RFI Operating life Environmental factors	

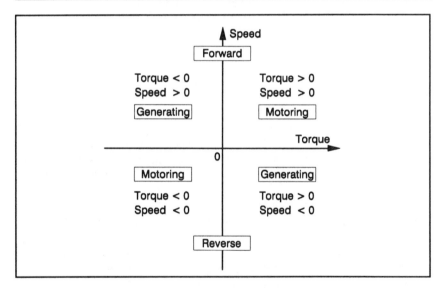

Fig. 10.1   Four quadrants of speed/torque diagram

10.3 Sizing the rotor; the Output Equation. Electric motor designs traditionally start with the *output equation*. In terms of the torque, the simplest form of this equation is

$$T = KD_r^2L \qquad [1]$$

where K is the *output coefficient* and D_r and L_{stk} are the *rotor* diameter and the stack length, respectively. K is proportional to the product of the electric and magnetic loadings, the definition of which requires a special treatment summarized in §10.9 and [67]. Typical values of K are given in the tables below.

	Typical values of σ, K, TRV		
	σ	K	TRV
	lbf/in^2	$lbf\text{-}in/in^3$	kNm/m^3
Small totally-enclosed motors	0.1 - 0.5	0.15 - 0.8	2.5 - 7
Integral-hp industrial motors	0.5 - 2	0.8 - 3	7 - 30
High-performance servomotors	1 - 3	1.5 - 5	15 - 50
Aerospace machines	2 - 5	3 - 7.5	30 - 75
Large liquid-cooled machines	10 - 15	15 - 200	100 - 250

	Guide values of σ
	$σ\ (lb/in^2)$
Low	< 1
Medium	1 - 2
High	> 2

Two other coefficients sometimes used for the same purpose are the *torque per unit rotor volume TRV* and the *airgap shear stess* σ. They are both proportional to K and the relationships between them are as follows.

$$TRV = \frac{T}{\frac{\pi}{4}D_r^2L}.$$ [2]

Therefore $TRV = 4K/\pi$. The airgap shear stress σ is the tangential (torque-producing) force per unit of swept rotor surface area:

$$TRV = 2\sigma.$$ [3]

Hence

$$K = \frac{\pi}{2}\sigma = \frac{\pi}{4}TRV$$ [4]

In this book the preferred coefficient is σ, measured in lbf/in^2. If D_r and L_{stk} are in inches, then T will be in lbf-in. Note that an airgap shear stress of $\sigma = 1$ lbf/in^2 corresponds to $TRV = 13.8$ kNm/m^3.

10.4 Envelope dimensions D_s and L_e. The *envelope dimensions* are the stator lamination diameter D_s and the axial length L_e measured over the end-turn overhangs, Fig. 10.2. They define a cylindrical volume which can be called the *gross electromagnetic volume*. The *net* electromagnetic volume is the cylindrical volume defined by D_s and L_{stk}. The electromagnetic volume does not include the frame, because there is so much variation in frame configuration.

10.5 Sizing the envelope dimensions when they are not specified. Suppose the torque requirement is specified; in the example, it is 10 lbf-in. Choosing $\sigma = 1$ lbf/in^2, $K = \pi/2 = 1.57$lbf-in/in^3 and the required rotor $D_r^2L_{stk}$ is obtained from equation 10.1 as $10/1.57 = 6.37$in^3. This fixes the rotor volume as 5.00in^3. To determine the rotor length and diameter separately, it is necessary to select the length/diameter ratio L_{stk}/D_r. A typical value is 1, so that $D_r^2L_{stk} = D_r^3 = 6.37$in^3 and $D_r = 1.85$in. Hence $L_{stk} = 1.85$in also. A longitudinal cross-section of this motor is shown in Fig. 10.2.

The simplest way to estimate the stator diameter D_s is from a typical or "standard" ratio of D_r/D_s. This ratio can vary over quite a wide range between 0.4 and 0.7, with most designs around 0.5-0.55. It depends on the number of stator and rotor poles, and on the operating requirements. A few *suggested* values are given in the table

Page 164

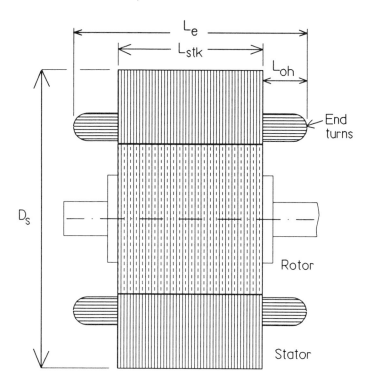

Fig. 10.2   Envelope dimensions L_e and D_s defining gross electromagnetic volume. This figure is a scale cross-section of the example motor.

below for the more common pole numbers. Note that the larger the number of poles, the larger the value of the ratio D_r/D_s tends to be. In the example $D_r/D_s = 0.5$ so $D_s = 1.85/0.5 = 3.7$in.

Phases	N_s	N_r	D_r/D_s	β_r	β_s
3	6	4	0.5	30°	32°
3	12	8	0.57	16°	15°
4	8	6	0.53	23°	21°

The overall length L_e is made up of the stack length L_{stk} plus two end-turn overhangs, one at each end:

$$L_e = L_{stk} + 2L_{oh}. \qquad [5]$$

L_e is an envelope dimension. The overhang length L_{oh} at each end can be estimated roughly as $1.2t_s$, where t_s is the stator tooth width (equation [7]). Hence

$$L_e \approx L_{stk} + 2.4\, t_s. \qquad [6]$$

Using a stator pole-arc of $\beta_s = 30°$ and an airgap length of 0.01in we get $t_s = 0.483$in and $L_{oh} = 0.58$in, so $L_e = 3.00$in. The envelope dimensions are therefore $D_s = 3.34$in and $L_e = 3.00$in. In Fig. 10.2 all these dimensions are drawn to scale. Note that the rotor is "square", with $L_{stk}/D_r = 1$. Values up to twice this value can be used. If L_{stk}/D_r is less than 1, end-effects begin to reduce the inductance ratio.

10.6 Sizing other internal dimensions.

Airgap g

The switched reluctance motor needs a uniform airgap and good concentricity to maintain balanced phase currents and minimize acoustic noise. It also requires a small airgap to maximize the specific torque and minimize the volt-ampere requirement in the controller. If the airgap is less than 0.25 mm (0.010 in) it may be necessary to grind or machine the stator bore or the rotor surface, or both, on the same centres as the bearing housings after the end-caps have been assembled to the stator. It may have been necessary to machine the stator OD to get an accurate register for the frame, and usually the stator stack will have been welded before that. With these steps, the airgap can be made as small as 0.1mm or 0.004in. Such small values are used only in extreme cases, although they are regularly used in stepper motors.

A rough guide to help choose the airgap length is 0.5% of the rotor diameter if the stator length/diameter ratio L_{stk}/D_r is 1, increasing in proportion to L_{stk}/D_r. For example, if $L_{stk}/D_r = 2$, the airgap should be increased to $0.01D_r$. In the example, $g = 0.005D_r = 0.005 \times 1.85 = 0.009$in.

Pole arcs β_s and β_r

β_s and β_r are constrained as discussed in Chapter 3:

(i) $\beta_r \geq \beta_s$. In terms of the widths, t_r should exceed t_s by between g and $2g$. This provides a slightly larger slot area without sacrificing aligned inductance. The difference in pole-widths increases increases the gap permeance coefficient from approximately $(t+g)/g$ for aligned poles of equal width, to $(t + 2g)/g$ when the narrower poles width is t, provided that the difference is more than about $2g$.

(ii) $\text{Min}(\beta_r, \beta_s) \geq \varepsilon$, the *stroke angle* or *step angle*. This ensures that in the ideal case with no fringing flux, torque can be produced at all rotor positions (provided that the pole and phase numbers are valid). It stems from the ideal linear theory in which one phase can produce torque over an angle equal to the stator or rotor pole arc, whichever is less. In practice the pole-arcs can be made less than the stroke angle; but by how much depends on the effective overlap ratio ρ_E (Chapter 3). Four-phase motors tend to have higher values of ρ_E and therefore the poles can be made narrower than ε without introducing torque dips; however, narrow poles reduce the aligned inductance and the inductance ratio.

(iii) $\beta_s < 2\pi/N_r - \beta_r$. This ensures that in the unaligned position, there is clearance between rotor and stator poles. The angular clearance between the stator and rotor pole-corners in the unaligned position is $\pi/N_r - \beta_r$, and this needs to be several degrees to ensure a sufficiently low unaligned inductance.

Once the pole-arcs have been decided, the pole-widths t_s and t_r follow from the equations

$$t_s = 2(r_1 + g)\sin\frac{\beta_s}{2} \qquad [7]$$

and

$$t_r = 2r_1 \sin\frac{\beta_r}{2}. \qquad [8]$$

Suggested pole-arcs have been given above for some of the more common switched reluctance motors. In the example, β_s = 30° and β_r = 32°. Then t_s = 0.483in and t_r = 0.510in.

Rotor slot depth d_r

The rotor slot depth is

$$d_r = r_1 - r_0 \qquad [9]$$

where the radius $r_1 = D_r/2$ and r_0 is sometimes called the *minor radius* of the rotor. The rotor slot depth should be at least 20-30 times the airgap length in order to obtain a low unaligned inductance. A useful guide is to make

$$d_r = \frac{t_s}{2}. \qquad [10]$$

However, there is no point in making the rotor slot too deep because, in the unaligned position, the flux from the stator pole tends to fringe into the sides of the rotor poles. If the angular clearance between the corners of the rotor and stator poles in the unaligned position is too small, deepening the rotor slot has no benefit in terms of reduced unaligned inductance. The depth of the rotor slot is also constrained by the need to make the rotor yoke thick enough to carry the peak flux without saturating, and also by the need to make the shaft diameter as large as possible.

In the example, equation [10] gives d_r = 0.483/2 = 0.242in.

Rotor yoke thickness y_r

The rotor yoke thickness y_r should be sufficient to carry the *peak* rotor flux without saturating. In a switched reluctance motor with a 2-pole flux pattern the main flux divides into two equal parts when it leaves the rotor pole and enters the rotor yoke. The yoke thickness y_r should be at least $t_r/2$, and preferably 20-40% more than this, to allow for the fact that sections of the yoke are shared between different phases which may overlap. In the example $y_r = (2/3)t_r = (2/3) \times 0.510 = 0.340$in.

Shaft diameter D_{sh}

A large shaft diameter is desirable to maximize the lateral stiffness of the rotor. This helps to minimize acoustic noise and to raise the first critical speed. An approximate formula for first critical speed is

$$n_c = 1.55 \times 10^6 \frac{d^2}{l \sqrt{Wl}} \qquad [11]$$

where l is the length between bearings [in], W is the rotor weight [lb], assumed to be concentrated at the centre of the shaft, and d is the shaft diameter [in][73].[1] In the example the shaft diameter has already been determined by choosing all the electromagnetic dimensions of the rotor, giving

$$
\begin{aligned}
D_{sh} &= D_r - 2(d_r + y_r) \\
&= 1.85 - 2(0.242 + 0.340) \qquad [12] \\
&= 0.686 \text{ in.}
\end{aligned}
$$

This value will be retained. If the length between bearings is taken to be $l = 2L_{stk} = 3.7\text{in}$, then with $d = 0.686\text{in}$ and $W = 0.774\text{lb}$ (calculated from the cross-section of the rotor), the first critical speed is calculated as 116,495rpm, which is well above the speed range for most applications.

Stator yoke thickness y_s; ovalisation

The stator yoke thickness y_s is subject to the same constraint as y_r: if t_s is the stator pole-width, then $y_s > t_s/2$, and preferably 20-40% more than this, to allow for the fact that sections of the yoke are shared between different phases which may overlap. The stator yoke sections are longer than the rotor yoke sections, so it is important to provide extra yoke thickness if possible. In the example, $y_s = (2/3)t_s$ = (2/3) × 0.483 = 0.322in.

---

[1] This formula assumes self-aligning bearings and takes no account of the radial compliance of the bearings or their housings; in other words, the shaft is assumed to be journalled in bearings that are infinitely stiff in the radial direction. A high critical speed does not mean that the rotor need not be balanced or that there will be no problems with vibration, but merely indicates whether the shaft is adequately stiff for the intended speed and rotor weight.

A further reason for making the stator yoke thicker is to maximize the stiffness of the stator against ovalising forces. This is important in reducing acoustic noise, and a yoke thickness $y_s = t_r$ would not be excessive from this point of view, although it would increase the weight and decrease the slot area. Any increase in yoke section that increases the stiffness against ovalising forces is desirable. The main methods for achieving a stiff yoke are:

(a) use the thickest possible yoke section;
(b) use a hexagonal or square lamination;
(c) use fillet radii in the slot corners;
(d) use tapered poles;
(e) compress the laminations tightly in the axial direction;
(f) use a stiff frame with an interference fit.

With two concentrated, diametrically opposite radial loads the diametral deflection of a plain ring is given by[2]

$$\Delta \approx \frac{1.8F}{E\left(\dfrac{t}{R}\right)^3} \qquad [13]$$

where E is the modulus of elasticity, F is the radial force per unit axial length, t is the radial thickness of the ring, and R is its mean radius. Clearly a small increase in the ratio t/R produces a significant gain in lateral stiffness. In the example the radial force can be calculated assuming that the rotor is in the aligned position with a flux-density of 2T in the poles. The magnetic force of attraction acting radially on the pole-faces is B^2/μ_0 per square meter of pole-face area. At 2T this is approximately 230lbf/in^2. In the example the pole-widths are about 0.5in, so the radial force per unit of axial length is $F = 115\text{lbf/in}$. The mean radius of the stator yoke is 1.69in, and if $E = 30 \times 10^6 \text{ lbf/in}^2$ the deflection from equation [13] is approximately 0.001in on the diameter. This is more than enough to cause considerable noise and vibration, especially if the force is cyclic and has a steep-fronted waveform. This estimate is, however, somewhat pessimistic because the flux-density in the aligned position will not normally be as high as 2T. In fact, *PC-SRD* calculates a peak

---

[2]Dr. P. Bhatt (Glasgow University), private communication. See also Timoshenko, *Strength of Materials*, Van Nostrand 1955, and Roark, *Formulas for Stress and Strain.*

Page 170

flux-density averaged over the stator poles of 1.57T, which reduces the force to 62% of the above value. Even so, a density of 2T or more can be experienced briefly when the poles are partially overlapped, perhaps 30% overlapped, in which case the radial forces will be only 1/3 as much. The calculation shows that this motor can hardly be considered to be inherently quiet. In an AC or brushless DC motor the radial forces are distributed over a much wider pole area and are less likely to cause ovalization of this order. By the same argument, switched reluctance motors that use short flux-paths should be inherently less susceptible to ovalization.

If possible the laminations should be held tightly in compression in the axial direction, and for best results they should be bonded, welded at three or four points on the OD, and mechanically compressed by a permanent clamping arrangement, because any tendency towards buckling will amplify the deflections and acoustic noise. Of course, all of these measures add cost.

Stator slot depth d_s

The stator slot depth needs to be as large as possible to maximize the winding area, making it easy to insert enough copper to minimize the copper losses. This is especially important for totally enclosed machines.

In the example d_s has already been determined by choosing all the other electromagnetic dimensions of the stator, giving

$$d_s = \frac{1}{2} (D_s - D_r - 2(g + y_s))$$

$$= \frac{1}{2} (3.70 - 1.85 - 2(0.009 + 0.322)) \qquad [14]$$

$$= 0.594 \text{ in.}$$

The resulting slot area is $A_{slot} = 0.479\text{in}^2$.

All of the geometric dimensions have now been determined, and they are summarized in the table overleaf.

PC-SRD's calculation of this motor at 2,000rpm is given on p.76. The actual torque (11lbf-in) is slightly higher than the specification.

Summary of example motor design			
Parameter	Symbol	Value	Units
Stator diameter	D_s	3.7	in
Rotor diameter	D_r	1.85	in
Stack length	L_{stk}	1.85	in
Overall length	L_e	3.00	in
Airgap	g	0.009	in
No. of stator poles	N_s	6	
No. of rotor poles	N_r	4	
No. of phases	m	3	
Stator pole arc	β_s	30	°
Rotor pole arc	β_r	32	°
Stator pole width	t_s	0.483	in
Rotor pole width	t_r	0.510	in
Stator slot depth	d_s	0.594	in
Rotor slot depth	d_r	0.242	in
Stator yoke thickness	y_s	0.322	in
Rotor yoke thickness	y_r	0.340	in
Shaft diameter	D_{sh}	0.686	in
Slot area	A_{slot}	0.479	in^2

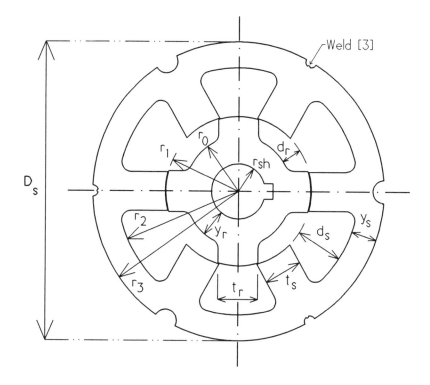

Fig. 10.3 Motor dimensions. This figure is a scale cross-section of the example motor.

10.7 The difficulty of hand calculation. The *peak* torque between overlapping poles can be estimated from Prof. Harris' formula, equation [4.27]. In the example motor, *PC-SRD* shows that the ratio of peak/mean torque is approximately 2.25 at the specification speed of 2,000rpm. This indicates that the peak torque must be about $2.25 \times 11 = 24.75$lbf-in. Then with $B_s = 1.7$T, Harris' formula gives the required peak ampere-turns/pole as

$$N_p i_{peak} = \frac{T_{peak}}{B_s D_r L_{stk}}$$

[15]

$$= \frac{24.75}{1.7 \times 1.85 \times 1.85} \times 175 = 744 \text{ A-t}$$

where the factor 175 is needed to convert the English units into metric units.[3] *PC-SRD* shows that at this operating point the peak/RMS current ratio is also about 2.25, so the RMS ampere-turns per pole is 744/2.25 = 331. Assume that the slot-fill factor is 33%. With two coil-sides per slot, the cross-sectional area of one coil-side (N_p turns) is 0.33 × 0.479/2 = 0.079in^2, and this gives an RMS *current-density* of 331/0.079 = 4,188A/in^2 (6.5A/mm^2). The peak current calculated by *PC-SRD* on p.76 is 26.2A, giving a peak ampere-turns/pole of 812, 9% higher than Harris' prediction, but the *PC-SRD* output shows a shaft torque of 11lbf-in, which when corrected for iron loss indicates an electromagnetic torque of 11.4lbf-in, 14% higher than Harris' formula. These comparisons provide a little extra confidence in the design but they also show the difficulty of "hand calculations" of the switched reluctance motor: they depend on ratios (peak/mean torque, peak/RMS current) which can be determined only by computer simulation.

The load-factor of the motor in the example is only 50%, so the true RMS current-density is only $(1/\sqrt{2}) \times 4{,}188 = 2{,}960$A/in^2 (4.6A/mm^2). This is a suitable value for a small switched reluctance motor but it would be wise to check it by means of a more detailed thermal calculation. With adequate cooling, higher values could be used.

The current-density calculation can be summarized in one equation:

$$J_{rms} \approx \frac{2 \times 175 \; k_{Tpeak} T}{k_{Ipeak} k_s A_{slot} B_s D_r L_{stk}} \text{ A/in}^2$$

[16]

where T is the mean torque [lbf-in]; $k_{T\,peak}$ is the ratio of peak to mean torque; $k_{I\,peak}$ is the ratio of peak to RMS current; k_s is the slot-fill factor (ratio of copper area to slot area); A_{slot} is the slot area [in^2]; B_s is the saturation flux-density [T]; D_r is the rotor diameter

---

[3] 1in = 1/39.37m and 1lbf-in = 1/8.85Nm, so the factor is $(39.37)^2/8.85 = 175$.

[in]; and L_{stk} is the stack length. (The formula assumes 2 coil-sides/slot). The formula can be simplified if it is assumed that $k_{T\ peak}$ and $k_{I\ peak}$ are equal, although they vary widely over the speed range. In particular, $k_{T\ peak}$ decreases at lower speeds.

If the current-density is too high, the slot area can be increased by increasing the stator slot depth but the stator OD may need to be increased at the same time. For totally enclosed motors the *continuous* current-density should be limited to about 2,500A/in² (4A/mm²). In domestic appliances higher values up to 6,000A/in² (10A/mm²) may be tolerable, and aerospace machines with oil-cooled windings can use as much as 12,000A/in² (20A/mm²).

10.8 Estimating the number of turns per pole N_p.

A rough estimate of the number of N_p can be made by assuming that at the specified speed the conduction angle of the power transistors has a certain value Δ; for example, $1/m$ of a rotor pole-pitch, which is equal to the stroke angle ε. If there is no current-chopping the peak flux-linkage per phase ψ_{peak} is given by

$$\psi_{peak} = \frac{V_s \Delta}{\omega} \qquad [17]$$

where ω is the angular velocity (= rpm×2π/60), V_s is the DC supply voltage, and Δ is in radians. At full speed ψ_{peak} occurs well before the aligned position, typically when the overlap between the stator and rotor poles is about 2/3 of the stator pole arc. At this point we assume that the ampere-turns are sufficient to bring the stator pole to the flux-density B_s across the entire width at its root. Then

$$\psi_{peak} = t_s L_{stk} B_s \times 2\,N_p \quad \text{V–s} \qquad [18]$$

Combining equations [18] and [19],

$$N_p = \frac{46,500\,V_s}{\text{rpm} \times m\,t_s\,L_{stk}\,N_r\,B_s} \qquad [19]$$

where t_s and L_{stk} are in inches and B_s is in T.[4] For the example motor with V_s = 24V, this gives N_p = 31 turns/pole if B_s = 1.7T.

---

[4]The factor 46,500 arises from the conversion of units.

10.9 Output coefficient; magnetic gear ratio. The output equation [1] can be used to analyze and compare the torque of the switched reluctance motor with that of AC motors, in particular the induction motor, by analyzing the output coefficient K in terms of the electric and magnetic "loadings" and the degree of utilization of these loadings in the two types of machine. Intuitively the electric and magnetic loadings give an idea of how much flux and how much current the machine is using.

In [67], Prof. M.R. Harris combined such an analysis with a detailed performance comparison between a switched reluctance motor and two induction motors (a standard-efficiency model and a high-efficiency model). The gist of his argument is paraphrased here, but the details of the comparison are too extensive to reproduce and reference should be made to the original paper. In passing, it is noted that surprisingly few detailed comparisons between switched reluctance motors and other types of machines have been published.

The classical definition of electric loading in an AC machine is

$$A = \frac{m N_{\text{ph}} I}{\pi D_{\text{r}}} \quad \text{A/m} \quad [20]$$

where I is the RMS phase current, m is the number of phases, D_{r} is the rotor diameter, and N_{ph} is the number of turns in series per phase. The magnetic loading B is considered as the *average* flux-density over the rotor surface. In AC motors the flux-density is distributed sinusoidally so that the fundamental flux per pole is given by

$$\Phi_1 = \text{B} \times \frac{\pi D_{\text{r}} L}{2p} \quad [21]$$

and then the generated EMF per phase is

$$E_1 = \frac{1}{\sqrt{2}} \omega k_{\text{w1}} N_{\text{ph}} \Phi_1 \quad \text{V} \quad [22]$$

The airgap power is $m E_1 I = T\omega/p$, and from these equations the torque per unit rotor volume can be determined as

$$\frac{T}{V_r} = KBA \quad \text{Nm/m}^3. \qquad [23]$$

The factor K includes an efficiency parameter to account for the power loss occurring between the airgap and the shaft, and a power-factor to account for the phase angle between E_1 and I, and in integral horsepower induction motors a value around 1.5-1.7 can be expected.

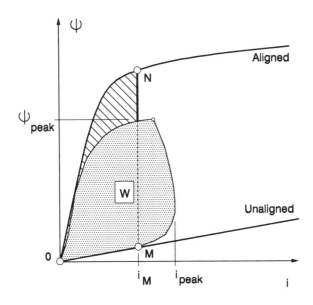

Fig. 10.4   Energy-conversion areas

For the switched reluctance motor Harris introduced the idea of the *time-averaged flux-linkage per phase* ψ_{ph}, which permits the magnetic loading to be defined in the same way as for the AC motor:

$$m\psi_{ph} = B\frac{\pi D_r L}{2p}N_{ph} \qquad [24]$$

where p is the number of pole-pairs in the basic flux pattern. Note in this equation that ψ_{ph} has to be multiplied by the phase number m

to get the total flux in the machine. This is a key idea, because it reflects the fact that in the switched reluctance motor the phases have their own fluxes and do not share a single main flux. ψ_{ph} is related to the peak flux-linkage per phase, ψ_{peak}, by the factor $K_\psi = \psi_{peak}/\psi_{ph}$, and Harris quotes a value of 2.77 for the motor discussed in [67].

The mean torque is obtained from the loop area W in the energy-conversion diagram (equation [2.12]):

$$T = \frac{mN_r}{2\pi}W \qquad [25]$$

Harris relates the area W to the "base" area $OMNO$ in Fig. 10.4, which is the energy-conversion area traced out by a square current waveform of peak value i_M and RMS value $I = i_M/\sqrt{2}$, where I is the same value of RMS current as in the *actual* current waveform. The base area $OMNO$ is unique and represents the maximum energy conversion (torque) available without exceeding the current i_M (peak). The actual loop area W has the peak current $i_{peak} > i_M$, but its on-time is shorter so that the RMS value is the same as that of the square waveform. The ratio

$$K_r = \frac{W}{OMNO} \qquad [26]$$

is defined, and Harris quotes a value of 0.858 and a typical range of 0.8-1.1. Next, the area $OMNO$ is related to the product $\psi_{peak}I$ by a second ratio

$$K_s = \frac{OMNO}{\psi_{peak}I}. \qquad [27]$$

The value quoted is 1.27. The electric loading is defined as for the AC motor:

$$A = \frac{2mN_{ph}I}{\pi D_r}. \qquad [28]$$

Combining equations [24]-[28], the torque per unit rotor volume is

$$\frac{T}{V_r} = GK_r K_s K_\psi \frac{1}{m}BA, \qquad [29]$$

where

$$G = \frac{N_r}{2p}.$$ [30]

G is the ratio of the angular velocity of the stator excitation pattern to the angular velocity of the rotor, and is termed the *magnetic gear ratio*. It is a characteristic of vernier reluctance motors [75], and it has the value 3 for a 4-phase 8/6 motor. Harris states that G raises the torque to an attractive level, "overcoming what would otherwise be a serious electromagnetic disadvantage". He explains this by pointing out that the phase flux-linkage is derived not from the whole magnetic loading (as in the AC motor) but from B/m (equation [29]). Each phase winding is extremely short-pitched, embracing only a fraction 1/m of what would be the total flux if all of the magnetic loading were available to each phase. Put another way, the switched reluctance motor makes poor utilization of the available periphery of the rotor. It compensates by executing a larger number of energy-conversion loops per revolution than there are pole-pairs in the basic flux pattern. If the induction motor were analyzed in terms of the area traced out on its ψ-i diagram, the loop would be elliptical and would have contributions from all four quadrants, as a result of the mutual inductance terms between phases.

Not only is the utilization of B low in the switched reluctance motor, but in the motor discussed in [67], the magnetic loading B is itself only half that of a high-efficiency induction motor. The electric loading is more than double, so that the product BA is 10% higher for the switched reluctance motor. When all the factors are put together, the torque per unit rotor volume is 62% higher for the switched reluctance motor, and the torque per unit stator volume (net electromagnetic volume) is 31% higher.

The higher power density of the switched reluctance motor in [67] (an early *OULTON*[5] motor) is achieved at the expense of a higher electric loading than that of the induction motor. To some extent this is achieved by making use of the large available conductor area, but the winding temperature rise is higher: 90°C against 38°C in the high-efficiency induction motor. However, this rise is still well within the

---

[5]*OULTON* is a trademark of Graseby Controls, Lowestoft, England

the permitted Class F limit. It is perhaps also worth noting that the *OULTON* motor was found to have a very high inductance ratio, reflecting the excellence of its electromagnetic design. At rated load the energy-conversion loop is a relatively small fraction of the available conversion area (as in Fig. 10.4) indicating the existence of a very large temporary overload capacity.

10.10 Calculation of volt-ampere requirement. The theory of the volt-ampere requirement of the switched reluctance motor is developed in [16], based on the naturally flat-topped current waveform associated with "base speed"; and Ray *et al* [28] have provided a commentary and data indicating that the switched reluctance motor has a slight advantage over the induction motor. Both papers quote values for the *specific peak volt-amperes*, defined as the product of the number of transistors times the peak current times the supply voltage, divided by the power conversion. The *RMS specific volt-amperes* is defined similarly, but using RMS current instead of peak current. For both the switched reluctance motor and the induction motor the specific peak volt-amperes is approximately 10kVA/kW. The theoretical comparison serves mainly to prove that the switched reluctance motor is not at a disadvantage relative to the induction motor, because the effect of saturation of the overlapping poles decreases the voltage requirement in the switched reluctance motor by a significant factor. If there were no saturation, the volt-ampere requirement of the switched reluctance motor would be excessively high [16]. Once this fact is established, the actual volt-ampere requirement must be calculated at all speeds and loads, and indeed for every individual power transistor and diode in the controller. For this purpose CAD software such as *PC-SRD* is essential.

11 TESTS AND MEASUREMENTS

11.1 Dimensions.

Airgap length g - Dimensional measurements on the switched reluctance motor are straightforward but special attention should be paid to the measurement of the airgap. The performance is sensitive to g in a number of ways which have been discussed already, but it is also sensitive to *variations* in g. In mass-produced motors, variations of 30% may be tolerated, but the switched reluctance motor needs tighter tolerances to ensure quiet operation and symmetrical loading of the phases.

The airgap length should be measured with feeler gauges if possible; otherwise the best alternative is to gauge the internal diameters between diametrically opposite poles using a dial-gauge with the stator set up on vee-blocks on a metrological table, and then to gauge the outside diameters across diametrically opposite rotor poles, Fig. 11.1. This method does not take account of the position of the shaft centreline in the assembled motor, but it does measure ovality of the stator and rotor. The measurements should be made at three or more positions along the length to check for skew or other distortions.

Overall length - It is useful to measure the overall length over the end-turn overhangs, L_e in Fig. 10.2. This helps build up experience on the allowance that must be made during design. It is also useful to know the extent of the end-turn overhangs when checking the measured winding resistance against the calculated value.

Weights - It is useful to measure the weights of the individual components as the motor is assembled (or disassembled), particularly the weight of the laminations, the copper, and the sum of the weights of the remaining parts (shaft, bearings, and frame). The iron and copper weights are useful in checking design calculations, and the copper weight, together with the bare wire diameter, is especially useful in checking the calculation of winding resistance.

Polar moment of inertia - The moment of inertia can be measured by means of a calibrated torsional pendulum. The rotor is suspended from the wire in a clamp that clasps the end of the shaft, Fig. 11.2. Care should be taken to machine this clamp so that the rotor axis is collinear with the line of the wire.

Fig. 11.1   Measuring the switched reluctance motor rotor

The number of swings in a timed period gives the torsional oscillation frequency f. Then the rotor is replaced by a reference inertia J_0 that fits the same clamp. If the frequency with J_0 is f_0, then the inertia of the switched reluctance motor is given by

$$J = J_0 (f_0/f)^2 \qquad [1]$$

If J_0 is a simple cylinder of mass M and radius r,

$$J_0 = \frac{Mr^2}{2} \quad \text{kg-m}^2 \qquad [2]$$

The moment of inertia is inherently low, but there is little published evidence that the torque/inertia *ratio* is any higher than that of specialized permanent-magnet brushless servo motors. Compared with induction-motor or DC-motor servos, the switched reluctance motor should have a higher T/J ratio. This does not mean that the switched reluctance motor is a good servo motor, however. The torque smoothness, the control linearity, and the ability to operate at and through zero speed are inherently worse than those of pure AC and DC motors, and although this disadvantage can in theory be overcome by current-profiling and other exotic electronic techniques, the justification for doing so is probably limited. For actuation and servo applications it is better to let the switched reluctance motor find its natural market in special niches - high-temperature environments, for example, and applications requiring a high degree of "fault tolerance", where PM motors are precluded or where induction motors would be too lossy (especially the rotors); or where very high speeds are required.

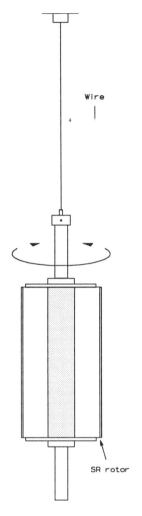

Fig. 11.2   Torsional pendulum

The coal-cutting machine in Fig. 9.8 is an interesting example in which low inertia is an advantage *per se*. If the cutters strike hard rock the whole machine stops suddenly, and the low inertia of the drive motor ensures that no damage can result to the drive shafts or to the motor itself.

11.2 Winding resistance. The simplest way to measure winding resistance is to use a multimeter. Greater accuracy is obtained with a four-terminal measurement, which is an option with many laboratory-type multimeters. Alternatively, a Wheatstone bridge can be used. The winding temperature should always be measured at the same time as the resistance is measured.

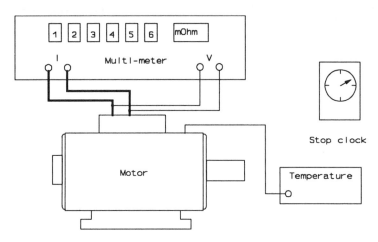

Fig. 11.3   Winding resistance measurement by 4-terminal method

The question sometimes arises as to the AC resistance of the switched reluctance motor, because the windings are subject to pulsed currents with high-frequency harmonic content. With non-sinusoidal waveforms the AC winding resistance is not easy to define or measure. When the current varies, the voltage across the winding terminals includes not only the Ri voltage drop and induced EMF, but also the referred EMF's induced in all conducting circuits that are coupled to the winding. These circuits include not only the other phases but also conducting paths in the laminated stator and rotor and in the frame. Moreover, the mutual coupling between all these circuits is a function of rotor position, like the self-inductance of the phase winding. In principle it is possible to conduct a sinusoidal frequency-response test and measure the in-phase and quadrature components of the terminal voltage, with the terminal current as reference phasor, over a range of frequency. This measurement includes the combined effect of all the coupled circuits, and gives an idea of the variation of impedance with frequency.

Unfortunately any variation of resistance with frequency will be accompanied by a reactive component which calls into question the validity of the simple circuit model of the winding. A more detailed analysis would have to include separate coupled circuits for all the conducting paths including those in the laminations. An alternative way to determine whether AC resistance effects are significant is to measure the static impedance at a fixed frequency that is typical of the highest frequency in the spectrum of the winding currents during normal operation, and then to repeat the measurement with a different winding having a different wire gauge, preferably with strands paralleled in such a way that the total conductor cross-section is maintained constant.

A rough estimate of the significance of AC resistance effects can be made in the following way. Suppose the supply voltage is V_s and the unaligned inductance is L_u. Then the maximum di/dt is V_s/L_u. To relate this to the conventional *skin-depth*, consider this value of di/dt as the peak di/dt of a sinewave current $i_{max} \sin \omega t$. Then

$$\omega = \frac{V_s}{L_u\, i_{max}}. \tag{3}$$

The skin-depth at this frequency is given by

$$\delta = \sqrt{\frac{2\rho}{\omega \mu_0}} \tag{4}$$

where $\rho = 1.7 \times 10^{-8}$ Ohm-m for copper at 20°C. In the example motor in Chapter 10, $V_s = 24$V, $i_{max} = 26$A, $L_u = 0.58$mH, and

$$\omega = \frac{24}{0.58 \times 10^{-3} \times 26} = 1,592 \text{ rad/s}. \tag{5}$$

The skin-depth in an isolated conductor at this frequency is

$$\delta = \sqrt{\frac{2 \times 1.7 \times 10^{-8}}{1,592 \times 4\pi \times 10^{-7}}} = 4.1 \text{ mm}. \tag{6}$$

The wire size in this motor is 1.45mm or 0.35δ, so skin-effect should not be significant. The proximity of iron parts can enhance the skin-effect, but individual strands of wire would probably need to be very close to the iron for the skin-effect to become significant.

11.3 Temperature rise. It is important to distinguish between *rise by resistance* and *hot-spot temperature*. The switched reluctance motor can have higher-than-normal hot-spot temperatures because the coil-sides have large cross-sections. (Induction and brushless DC motor windings are more finely divided in a larger number of slots, providing better thermal contact and diffusion.) The hot-spot temperature can be measured by an embedded thermocouple in the winding. The rise by resistance is usually measured after a heat run. The motor is run at full load long enough to ensure that the temperature distribution and heat flows have stabilized. It is disconnected and a stop-watch is started at the same time, Fig. 11.3. The resistance is then measured at intervals with no forced cooling. Assuming an exponential decrease of resistance,

$$R = R_0 e^{-t/\tau} \qquad [7]$$

where R_0 is the resistance at time $t = 0$ and τ is the time constant. Taking logarithms to base e,

$$\ln R = \ln R_0 - \frac{t}{\tau} \qquad [8]$$

which is the equation of a straight line, plotting $\ln R$ vs. time t, Fig. 11.4. Backward extrapolation to $t = 0$ gives the value of $\ln R_0$ and hence R_0. The resistance varies with temperature according to

$$R = R_b (1 + \alpha \Delta T) \qquad [9]$$

where α is the temperature coefficient of resistance (0.00393 for copper), R_b is the base resistance, and ΔT is the temperature rise above the base temperature (normally room temperature): thus

$$\Delta T = \frac{R - R_b}{\alpha R_b}. \qquad [10]$$

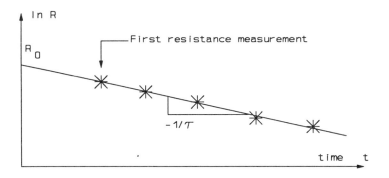

Fig. 11.4   Extraction of resistance at time $t = 0$.

11.4 Efficiency and Losses. Usually the efficiency that is most important to the user is the total system efficiency from the AC or DC supply through to the shaft: this includes the efficiencies of the motor and the controller:

$$\eta = \frac{P_m}{P_s}. \qquad [11]$$

The AC or DC supply is usually filtered from the worst effects of the pulsating or harmonic currents that circulate between the motor and the controller, so the input power at the source can be measured with a three-phase electronic wattmeter that samples at a relatively low frequency (e.g. a few kHz), Fig. 11.5.

Separation of the motor and controller efficiencies requires the measurement of the power flow P_C from the controller to the motor:

$$\eta_{motor} = \frac{P_m}{P_C} \qquad [12]$$

and

$$\eta_{controller} = \frac{P_C}{P_s}. \qquad [13]$$

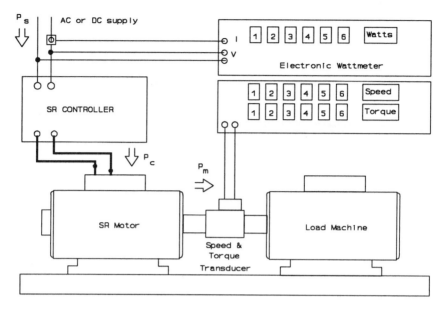

Fig. 11.5   Measurement of efficiency

P_c contains significantly more harmonics than either P_m or P_s, and its accurate measurement requires a sampling rate higher than is normally used by conventional electronic wattmeters. This can be understood as follows. The basic definition of *average power* is

$$P = \frac{1}{T} \int_0^T vi\, dt. \qquad [14]$$

In the case of P_s the voltage v is the supply voltage, which is normally stiff, or heavily filtered, or both: that is, for a DC supply

$$v = V_s \approx constant. \qquad [15]$$

The average power P_s is therefore approximately

$$P_s = \frac{V_s}{T} \int_0^T i\, dt; \qquad [16]$$

and the harmonics in P_s are therefore of the same order as those in the supply current i. In the case of P_C, the voltage v in equation [14] is the terminal voltage of the controller, which is rich in harmonics: in fact, it has much higher harmonic content than the current because of the filtering effect of the winding inductance. The harmonic power spectrum extends up to double the highest harmonic order in the winding current. Furthermore, the measurement of P_C requires separate measurements on the separate phases. Accurate measurement of P_C can be accomplished with a data acquisition system that records the separate waveforms of phase voltage and phase current. These waveforms are ported to a computer which executes equation [14] for all the phases separately, and adds them.

11.5 Static magnetization curves and static torque/angle curves. The static magnetization curves should be measured and compared with calculated values. They can be determined either directly, or from the static torque/angle curves [76].

The direct method for measuring magnetization curves depends on the voltage equation

$$v = Ri + \frac{d\psi}{dt} \qquad [17]$$

where v is the terminal voltage of the phase winding, R is its resistance, and i is the current. The magnetization curve measurement is based on the integration of this equation:

$$[\psi] = \int ([v] - [i]) \, dt \qquad [18]$$

where the square brackets indicate waveforms stored digitally as arrays of samples.

In general the phase flux-linkage ψ is a function not only of the current i, but also of currents in all the other phases *and* of induced currents in coupled conducting components such as laminations, shaft and frame. The magnetization curve measurement is conducted with all the other phases open-circuited; but the elimination of the effects of induced currents in coupled conducting paths requires more care. One way is to conduct the measurement with a sufficiently low

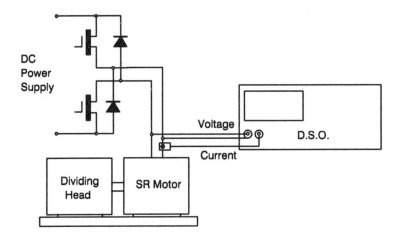

Fig. 11.6   Set-up for measuring magnetization curves

voltage so that the induced currents in these coupled paths are negligible. The other way is to use an adaptation of the Prescot bridge [77], which eliminates their effects absolutely.

Fig. 11.6 shows the experimental arrangement. The rotor is coupled to a machinist's dividing head which positions the rotor very precisely and holds it rigidly in position against the extremely high torques that can be experienced during the test. The phase winding is connected to a low-frequency square-wave voltage supply that can supply a peak current typically at least equal to the peak current experienced by the motor in normal operation. A convenient circuit for this purpose is the half-bridge or phaseleg circuit (one-third of the circuit in Fig. 6.1). The waveforms of terminal voltage and current are recorded on a digital processing oscilloscope or in a digital data acquisition system. Typical waveforms are shown in Fig. 11.7. The flux-linkage/current curve for a given position, [ψ] vs. [i], is obtained by digital integration of equation [18], and may be plotted in X-Y mode on the oscilloscope. The rotor is rotated to a set of equally-spaced positions and the magnetization curves for all these positions are computed, making up a complete set as shown in Fig. 2.2.

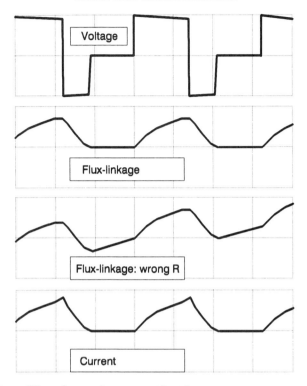

Fig. 11.7   Waveforms in magnetization curve measurement

A difficulty with this method is the choice of a suitable value for R (scaled in equation [18] to account for the voltage/current ratio of the current sensor). Unfortunately R changes with temperature as the winding heats up during the test. This effect can be minimized by operating the test on a single-shot basis, but the *effective* value of R is a function of time because it includes induced-current effects in the coupled conducting paths mentioned earlier. The effect of having the wrong value of R is to introduce a drift term into the flux-linkage waveform, Fig. 11.7. The value of R used in the digital integration is adjusted arbitrarily until the zero sections of the flux-linkage waveform are truly horizontal. Since $\psi = 0$ when $i = 0$, this correctly compensates for any variation in R for the reasons mentioned.

The magnetization curves can also be constructed from the *static torque curves*, which plot torque versus rotor position for a set of different currents. The procedure is described in [76].

Index

References

1. Nasar SA [1969] DC switched reluctance motor. Proceedings IEE, Vol 116, No 6, 1048-9

2. Lawrenson PJ et al [1980] Variable-speed switched reluctance motors, Proceedings IEE Vol 127 Pt B 253-265. See also Discussion, *ibid.*, 260-8

3. Miller TJE [1988] Switched reluctance motor drives. PCIM Reference Book, Intertec Communications, Inc., Ventura, California

4. Hendershot JR [1989 - US Patent]

5. Byrne, JV et al [1985] A high-performance variable reluctance motor drive: a new brushless servo. Motorcon Proceedings 147-60. See also H.M. Noad [1859] : A manual of electricity. (Lockwood, London)

6. Bedford BD [1972] US Patent Nos. 3678352 and 3679953

7. Unnewehr LE and Koch WH [1974] An axial air-gap reluctance motor for variable-speed applications. IEEE Transactions, PAS-93, 367-76

8. Bausch H and Rieke B [1978] Speed and torque control of thyristor-fed reluctance motors. Proceedings ICEM, Vienna Pt. I, 128.1-128.10. Also: Performance of thyristor-fed electric car reluctance machines. Proceedings ICEM, Brussels E4/2.1-2.10

9. Byrne JV and Lacy JG [1976] Characteristics of saturable stepper and reluctance motors. IEE Conf. Publ. No. 136, Small Electrical Machines, 93-6

10. Euxibie E and Thenaisie P [1990] A switched reluctance drive for pallet truck applications. Intelligent Motion, June Proceedings, 88-100

11. See also Maccon Machine Control No. 10 published by Maccon GmbH in Munich, describing the SR motor as a servomotor.

13. Regas KA and Kendig SD [1987] Step-motors that perform like servos. Machine Design, December 10, 116-120

14. Konecny K [1981] Motor-con Proceedings, 2A2.1-11

15. Ray WF and Davis RM [1979] Inverter drive for doubly salient reluctance motor; its fundamental behaviour, linear analysis, and cost implications. Electric Power Applications, 2, 185-93

16. Miller TJE [1985] Converter volt-ampere requirements of the switched reluctance motor drive. IEEE Transactions IA-21 1136-44.

17. Harris MR, Finch JW, Mallick JA and Miller TJE [1986] A review of the integral-horspeower switched reluctance drive. IEEE Transactions IA-22, 716-21.

18. Harris MR, Finch JW and Metwally HM : Proceedings IEE paper to be

published 1992 or 1993

19. Byrne JV US patent : controlled saturation

20. El-Khazendar MA and Stephenson JM [1986] Analysis and optimization of the 2-phase self-starting switched reluctance motor. ICEM, Munich, Sep. 8-10, 1031-1034

21. Compter JC [1984] Microprocessor-controlled single-phase reluctance motor. Drives/Motors/Controls, Brighton, 64-8

22. Horst G [1992] United States patent No. 5122697

23. Hutton AJ and Miller TJE [1991] Use of flux-screens in switched reluctance motors. IEE Fourth International Conference on Electrical Machines and Drives, 13-15 September, 312-316

24. Finch JW, Harris MR, Musoke A, and Metwally HMB [1984] Variable-speed drives using multi-tooth per pole switched reluctance motors. 13th Incremental Motion Control Systems Society Symposium, University of Illinois, Urbana-Champaign, IL, 293-302

25. Hendershot JR [1989] Short flux paths cool SR motors. Machine Design, 106-111

26. Hendershot JR [1991] A five-phase switched reluctance brushless DC motor with a low-loss magnetic circuit. Proceedings of the Incremental Motion Control Systems Society Symposium.

27. Byrne JV and O'Dwyer JB [1976] Saturable variable reluctance machine simulation using exponential functions. Proceedings of the International Conference on Stepping Motors and Systems, University of Leeds, July, 11-16

28. Ray WF, Lawrenson PJ, Davis RM, Stephenson JM, Fulton NN and Blake RJ [1986] High performance switched reluctance brushless drives. IEEE Transactions, Vol. IA-22, No. 4, 722-730

30. Michaelides AM and Pollock C [1992] A new magnetic flux pattern to improve the efficiency of the switched reluctance motor. IEEE IAS Annual Meeting, Houston, Texas. Conference Record, 226-233.

31. PC-SRD User's manual [1992] Version 5.2

32. Lovatt HC and Stephenson JM [1992] Influence of number of poles per phase in switched reluctance motors. IEE Proceedings-B, Vol. 139, No. 4, 307-314

33. Miller TJE and McGilp M [1990] Nonlinear theory of the switched reluctance motor for rapid computer-aided design, Proceedings IEE, 137, Pt. B, No. 6, 337-347.

34. Miller TJE and McGilp M [1991] High-speed CAD for brushless motor drives.

REFERENCES

European Power Electronics Conference, Florence, 3.435-439

35. Stephenson JM and Čorda J [1979] Computation of torque and current in doubly-salient reluctance motors from nonlinear magnetization data. Proceedings IEE, Vol. 126, No. 5, 393-6

36. Harris MR and Sykulski JK [1991] Simple method for calculating the peak torque of a switched reluctance motor: a computational investigation. Proceedings of the International Symposium on Electromagnetic Fields in Electrical Engineering, 18-20 September, Southampton, UK

37. Trowbridge CW [1990] An introduction to computer aided electromagnetic analysis. Published by Vector Fields Ltd., ISBN 0 9516262 0 5

38. Pollock C and Williams BW [1990] Power convertor circuits for switched reluctance motors with the minimum number of switches. IEE Proceedings-B, Vol. 137, No. 6

39. Miller TJE [1987] Brushless reluctance motor drives. IEE Power Engineering Journal, Vol. 1, 283-9

40. Krishnan R, Aravind S, and Materu P [1987] Computer-aided design of electrical machines for variable speed applications. Proceedings IEEE-IECON'87, Cambridge, 756-763

41. Stephenson JM and El-Khazendar MA [1989] Saturation in doubly salient reluctance motors. IEE Proceedings-B, Vol. 136, No. 1, 50-58

42. Vukosavic S and Stefanovic VR [1990] SRM inverter topologies: a comparative evaluation. IEEE IAS Annual Meeting, Conf. Record, Seattle, WA

43. Bass JT, Ehsani M, Miller TJE and Steigerwald, RL [1987] Development of a unipolar converter for switched reluctance motor drives. IEEE Transactions, Vol. IA-23, 545-553

44. Byrne JV, McMullin MF, and O'Dwyer JB [1985] A high-performance variable reluctance drive: a new brushless servo. Motor-Con Proceedings, October 147-160

45. Miller TJE, Cossar C, Anderson D [1990] A new control IC for switched reluctance motor drives. IEE Conference on Power Electronics and Variable Speed Drives, London, July 17-19, 331-335

46. Ray WF, Davis RM, and Blake RJ [1986] The control of SR motors. Conference on Applied Motion Control, Minneapolis June 137-145 (in [3])

47. Corda J and Stephenson JM [1982] Speed control of switched reluctance motors. International Conference on Electrical Machines, Budapest (in [3])

48. Chappell PH, Ray WF, and Blake RJ [1984] Microprocessor control of a variable reluctance motor. Proceedings IEE, Vol. 131, Pt. B., No. 2, 51-60

49. Miller TJE, Bower PG, Becerra R, and Ehsani M [1988] Four-quadrant brushless reluctance motor drive. IEE Conference on Power Electronics and Variable Speed Drives, London (in [3])

50. Bose BK, Miller TJE, Szczesny PM, and Bicknell, WH [1986] Microcomputer control of switched reluctance motor. IEEE Transactions, Vol. IA-22, 708-715

51. Ilic-Spong M, Miller TJE, MacMinn SR, and Thorp JS [1987] Instantaneous torque control of electric motor drives. IEEE Transactions, Vol. PE-2, 55-61

52. Lumsdaine, AH, Lang JH, Balas, MJ [1986] A state observer for variable reluctance motors. 15th Annual Incremental Motion Control Systems Society Symposium, University of Illinois, Urbana-Champaign, Illinois, June 267-273 (in [3])

53. Harris, WD and Lang, JH [1988] A simple motion estimator for VR motors. IEEE IAS Annual Meeting, Pittsburgh, PA, October 1988 (in [3])

54. Acarnley PP, Hill RJ and Hooper CW [1985] Detection of rotor position in stepping and switched reluctance motors by monitoring of current waveforms. IEEE Transactions, Vol. IE-32, No. 3, August 215-222 (in [3])

55. Bass JT, Ehsani M, and Miller TJE [1986] Robust torque control of a switched reluctance motor without a shaft position sensor. IEEE Transactions, Vol. IE-33, No. 33, 1986 212-216

56. Frus JR and Kuo BC [1976] Closed loop control of step motors without feedback encoders. Proceedings of the Fifth Annual Symposium on Incremental Motion Control Systems and Devices, May CC1-CC11

57. Kuo BC and Cassat A [1977] On current detection in variable reluctance step-motors. Proceedings of the Sixth Annual Symposium on Incremental Motion Control Systems and Devices, Urbana-Champaign, May 205-220

58. MacMinn SR, Szczesny PM, Rzesos WJ and Jahns TM [1988] Application of sensor integration techniques to switched reluctance motor drives.

59. Hedlund G and Lundberg H [1989] United States Patent No. 4,868,478

60. Hill RJ and Acarnley PP [1985] United States Patent No. 4,520,302

61. Ilic-Spong M, Marino R, Peresada SM, and Taylor DG [1987] Feedback linearizing control of switched reluctance motors. IEEE Transactions, Vol. AC-32, No. 5, 371-379

62. Wallace RS and Taylor DG [1991] Low torque ripple switched reluctance motors for direct-drive robotics. IEEE Transactions on Robotics and Automation, Vol. 7, No. 6, 733-742

63. Wallace RS and Taylor DG [1992] A balanced commutator for switched reluctance motors to reduce torque ripple. IEEE Transactions on Power

REFERENCES

Electronics, October 1992

64. McClelland ML, Lovatt HC and Stephenson JM [1991] Wide-bandwidth torque and power measurement in reluctance motor drives. European Power Electronics Conference, Florence, September 1991

65. Stephens CM [1989] Fault detection and management system for fault tolerant switched reluctance motor drives. IEEE IAS Ann. Mtg., Conf. Rec. 574-578

66. Miller TJE and Jahns TM [1986] A current-controlled switched reluctance drive for FHP applications. Conference on Applied Motion Control, Minneapolis, June 10-12 (in [3])

67. Harris MR and Miller TJE [1991] Comparison of design and performance parameters in switched reluctance and induction motors. IEE Fourth Internat. Conference on Electrical Machines and Drives, 13-15 September, 303-307

68. Materu P and Krishnan R [1988] Estimation of switched reluctance motor losses. IEEE IAS Meeting, Pittsburgh, PA, October (in [3])

69. Slemon GR and Liu X [1990] Core losses in permanent magnet motors. IEEE Transactions on Magnetics, Vol. 26, No. 5, 1653-1655

70. Holman JP [1989] Heat Transfer, McGraw-Hill, ISBN 0-07-100487-4

71. Hendershot JR [1991] Design of brushless permanent magnet motors. Magna Physics Publishing Division, Hillsboro, Ohio

72. Soong, WL and Miller TJE EMD 93

73. Miller TJE and Bower PG [1987] European Patent Application

74. Machinery's Handbook, 14th Edn., 1953

75. Mukherji KC and Tustin A [1974] Vernier reluctance motor. Procedings IEE, Vol. 121, No. 9, 965-974

76. Cossar C and Miller TJE [1992] Electromagnetic testing of switched reluctance motors. International Conference on Electrical Machines, Manchester, September 15-17, 470-474

77. Jones CV [1967] The unified theory of electrical machines. Butterworth & Co, London

78. Miller TJE [1989] Brushless reluctance and permanent-magnet motor drives. Oxford University Press

79. Lawrenson PJ [1992] A brief status review of switched reluctance drives. EPE Vol. 2, No. 3, 133-144

80. Lawrenson PJ [1992] Switched reluctance : a perspective. ICEM 92